Progress in Computer Science and Applied Logic
Volume 14

Learning and Geometry: Computational Approaches

David Kueker
Carl Smith
Editors

1996

Birkhäuser
Boston • Basel • Berlin

David W. Kueker
Department of Mathematics
University of Maryland
College Park, MD 20742

Carl H. Smith
Department of Computer Science
University of Maryland
College Park, MD 20742

Library of Congress Cataloging-in-Publication Data
Learning and Geometry / David Kueker, Carl Smith, editors.
p. cm. -- (Progress in computer science and applied logic ; v. 14)
Includes bibliographical references.
ISBN 0-8176-3825-3 (h) . -- ISBN 3-7643-3825-3 (h)
1. Computer vision--Congresses. 2. Geometry--Congresses.
I. Kueker, D. W., 1943- . II. Smith, Carl, 1950 Apr. 25-
III. Series.
TA 1634.L43 1966
006.3'7--dc20 95-41450
CIP

Printed on acid-free paper

Birkhäuser

ISBN 0-8176-3825-3
ISBN 3-7643-3825-3
Camera ready copy provided by the editors.
Printed and bound by Quinn-Woodbine, Woodbine, NJ.
Printed in the U.S.A.

9 8 7 6 5 4 3 2 1

Table of Contents

Preface

The field of computational learning theory arose out of the desire to formally understand the process of learning. As potential applications to artificial intelligence became apparent, the new field grew rapidly. The learning of geometric objects became a natural area of study. The possibility of using learning techniques to compensate for unsolvability provided an attraction for individuals with an immediate need to solve such difficult problems.

Researchers at the Center for Night Vision were interested in solving the problem of interpreting data produced by a variety of sensors. Current vision techniques, which have a strong geometric component, can be used to extract features. However, these techniques fall short of useful recognition of the sensed objects. One potential solution is to incorporate learning techniques into the geometric manipulation of sensor data. As a first step toward realizing such a solution, the Systems Research Center at the University of Maryland, in conjunction with the Center for Night Vision, hosted a Workshop on Learning and Geometry in January of 1991. Scholars in both fields came together to learn about each others' field and to look for common ground, with the ultimate goal of providing a new model of learning from geometrical examples that would be useful in computer vision.

The papers in the volume are a partial record of that meeting. In addition to the research papers submitted for these proceedings we include the program for the workshop, a list of participants, and the introductory remarks concerning the purpose of the workshop prepared by Vincent Mirelli of the Center for Night Vision.

David Kueker
Carl Smith

List of Participants

John S. Baras
Eric B. Baum
Jürgen Bokowski
Roger Brockett
John Case
Shang-hing Chou
Bob Daley
Mark Fulk
K.P. Jantke
George Jones
Michael Kim
Teresa Kipp
P.S. Krishnaprasad
David Kueker
Sanjeev R. Kulkarni
Josip Loncarlc
Vincent Mirelli
Jorma Rissanen
Kathleen Romanik
Haim Shvayster
Robert Sloan
Carl Smith
Vladmir Vapnik
Walter Whiteley
Rolf Wiehagen
Wu Wen-Tsun
Wlodek Zadrozny

Program for the Workshop

Tuesday, January 8, 1991

Plenary Talk, "Learning Patterns Based on Conditional Density Propagation," Roger Brockett, Harvard University

"Representing Geometric Configurations," Walter Whiteley, Champlain Regional College

"Mechanical Geometry Theorem Using Algbraic Methods," Shang-Ching Chou, University of Texas at Austin

Wednesday, January 9, 1991

Plenary Talk, "Some Aspects of Learning and Geometry in Computational Synthetic Geometry," Jurgen Bokowski, University of Darmstadt

"MDL Learning," Jorma Rissanen, IBM, Almaden Research Center

"Induction Principles in the Learning Theory," Vladimir Vapnik, Visitor, AT&T Bell Laboratories

Thursday, January 10, 1991

Plenary Talk, "Recursion Theoretic Learning Theory," John Case, University of Delaware

"Vague Predicates and Rules of Abduction," Wlodek Zadrozny, IBM Hawthorne

"'PAC' Learning from Noisy Data," Robert Sloan, University of Illinois at Chicago

Friday, January 11, 1991

Plenary Talk, "Geometry Theorem Proving in Euclidean, Descartesian, Hilbertian and Computerwise Fashions," Wen-Tsun Wu, Institute of System Sciences, China

"Naturalness in Inductive Inference," Rolf Wiehagen, der Humboldt Universität zu Berlin

"When Pluralistic Learners Behave Probabilistically for Finite Inductive Inference," Robert Daley, University of Pittsburgh

Introduction

Vincent Mirelli

Human vision can be interpreted as the function of learning the lowest complexity properties and invariants characterizing the useful partitions in classes of geometrical shapes (and of the geometrical components of complex figures). So it involves in an essential way geometry and learning. Thus, high-level human vision skills can be studied within the context of learning from geometrical examples. In turn, the learning process is related with the apparent ability of humans to establish and use a network of relationships between several different representations of classes of geometrical shapes at various levels classified as semantic or syntactic/symbolic/algebraic, depending on context.

Further progress in computer vision requires a careful reexamination of vision fundamentals along the interpretation described above, towards discovering the appropriate mathematical foundations and semantics which best fit vision problems. Towards this end we organized a cross disciplinary workshop at the Systems Research Center (now called the Institute for Systems Research) of the University of Maryland on "Learning and Geometry." The central objective of this workshop was to bring together several prominent mathematical researchers from fields which hold promise in establishing a solid mathematical framework linking learning and geometry in the context of vision. The topical fields represented in the conference are: computational linguistics, geometry theorem proving, synthetic, foundational and algebraic geometry. We provide below a brief description of the relevance and promise of each towards the central theme of this conference.

Computational learning [2], [11] addresses the problem of learning from examples. The Valiant model of learning [3] has been successful in several problems [4]. However, we believe that it needs to be extended and modified in order to address the specific human talents and skills involved in vision. The vision phenomenology is considerably more structured than previously addressed examples in the computational learning literature. Human exploits in a crucial way this additional structure and the interrelationships between model theoretic semantic structure and the mathematical representations in various formal systems. This type of interrelationships is analogous to the learning formation of natural language, where syntactical constraints are induced by the rules which map meanings into linear surfaces, and vice versa.

This interpretation makes it apparent that the field of *model-theoretic semantics* [5] is a natural context in which to formulate the understanding of these interrelationships. It is also appropriate for the organization of contributions from other topical fields represented in this conference. Model-theoretic semantics is at the heart of human learning, for instance in natural language learning. It is one of the goals of this conference to understand the role of model-theoretic semántics in vision. In the context of learning and geometry (in particular as it relates to vision) we need methodologies that can efficiently link different semantic and syntactic representations and related models, which focus on different views of the empirical data and provide the framework for logical intcrfcrcncc along different views (for instance, the local, continuous, global, discrete, algebraic, and symbolic aspects). In vision it is necessary to provide a framework for studying the linkages between these different representations and the ability to reason accross such semantic barriers. Model theoretic semantics and extentions can play that role.

The view expressed here for the mathematical foundations of vision is analogous to the view that has generated the modern field of *computational linguistics* in place of classical linguistics. From this point of view formal languages and *computational linguistics* [7] provide the mathematical framework for analyzing the structure of the various representations and the resulting syntactical constraints. In addition they provide the means for formal evaluation of different representations (e.g., expressive power of languages) and formal means for classifying languages and linking representations. Since learning ability (that is the concept of class of shapes being learnable) depends heavily on representation (i.e., the language used), and since we are interested in a framework that permits formulation of the concept of learnable based on concurrently linked representations, computational linguistics is seen as a significant methodological ingredient towards the overall theme of the conference.

The field of *geometry theorem proving* has been very successful in its goal [6], [9], while automatic theorem proving has not [8]. The reason is that automatic theorem proving tries to imitate the way human beings prove theorems, while geometry theorem proving is inspired by the relation between synthetic and analytic geometry. This relationship is analogous to that between syntax and semantics, which we believe is fundamental in the learning process. The additional relevance of this discipline on learning from geometrical example originates from the fact that there is a profound interpenetration among human talents of empirical learning, theorem proving, and calculus (i.e., chains of important algebraic operations performed almost without thinking). It is self evident, that this topical field, based on synthetic, foundational and algebraic geometry [1], [10], would also provide the formal framework to obtain the representations needed to characterize sets of geometrical shapes.

References

[1] *Fundamentals of Mathematics: Volume 2 Geometry*, H. Behnke, F. Bachman, K. Fladt, and H. Kunle, eds., MIT Press, Cambridge, MA, 1974.

[2] *Proceedings of the 1988 Workshop on Computational Learning Theory III*, Morgan Kaufman Publishers, 1988.

[3] L.G. Valiant, "A theory of the learnable," *Communications of the ACM*, **27**(11), Nov. 1984, pp. 1134–1142.

[4] H. Shvayster, "Learnable and nonlearnable visual concepts," *IEEE Transactions on Pattern Analysis and Machine Intelligence*, **12**(5), May 1990, pp. 459–466.

[5] C.C. Chang and H.J. Keisler, *Model Theory*, North Holland, 1973.

[6] *Geometric Reasoning*, D. Kapur and J.L. Mundy, eds, MIT Press, Cambridge, MA, 1988.

[7] *Computation of Language: an Essay on Syntax, Semantics and Pragmatics in Natural Man-Machine Communication*, R. Hausser, Springer-Verlag, New York, 1989.

[8] *The Artificial Intelligence Debate: False Starts and Real Foundations*, S.R. Graubard, ed., MIT Press, Cambridge, MA, 1988.

[9] S.-C. Chou, Mechanical geometry theorem proving, *Mathematics and its Applications,* **41**, D. Reidel Publishing Co., 1988.

[10] B. Sturmfels and J. Bokowski, *Computational Synthetic Geometry*, Springer-Verlag, 1989.

[11] M.J. Kearns, *The Computational Complexity of Machine Learning*, MIT Press, Cambrige, MA, 1990.

LEARNING

Learning by MDL

J. Rissanen
IBM Almaden Research Center
San Jose, Ca 95120-6099 USA

Bin Yu
Dept. of Statistics
University of California
Berkeley, Ca 94720 USA

Abstract

Machine learning has been formalized as the problem of estimating a conditional distribution as the 'concept' to be learned. The learning algorithm is based upon the MDL (Minimum Description Length) principle. The asymptotically optimal learning rate is determined for a typical example.

1. Introduction

In Valiant (1984) 'learning' was formalized as the process of estimating a Boolean function (or a related 'concept') in a certain class from randomly generated samples of the portion of its graph where the function has value 1. A special emphasis was put to the requirement that for learnability any desired error rate be achieved from a sample such that both its size and the required computation time are bounded by a polynomial function of the various parameters characterizing the error rate and the complexity of the function.

However, it seems that an adequate theory of learning is not possible without allowance for noise in the data, and after several approaches to generalize Valiant's framework, Angluin and Laird (1988), Kearns and Li (1988), Haussler (1989) and others, Yamanishi (1990) proposed the sensible idea that 'learning' is tantamount to estimating conditional distributions of the type $P(y|x)$, where y ranges over a finite set while the range of the 'features' x is essentially arbitrary. The data are assumed to be generated by independent sampling from a joint distribution $P(y,x)$ belonging to a family $\mathcal{P}$, but we are only interested in estimating the conditional distribution. Yamanishi studied in detail discrete features, such as those appearing in 'stochastic decision lists', which generalize the decision lists in Rivest (1987). He also generalized Valiant's requirements of polynomial bounds on the sample size and the computation time, and since any function, a Boolean function in particular, is a degenerate conditional distribution, this formulation of learning theory is strictly more general than Valiant's. Further, Yamanishi proposed a learning algorithm to be derived with the MDL principle, which he analyzed in detail. For a related study, see also Suzuki (1990).

While polynomial restrictions on the speed with which the estimation is to be done for a concept to be learnable are desirable from a practical point of view, we feel that they have nothing to do with learnability per se. After all, we certainly can extract useful information from data even if the entire object is too complex to be 'learned' in polynomial time, which to us appears to be the essence in all learning. If we wish to impose conditions for learnability, it seems that the requirement of consistency is the thing to ask. In broad terms, this amounts to the requirement that the estimated conditional distribution $\hat{P}(y|x,y^n,x^n) \rightarrow P(y|x)$ in a suitable sense, as $n \rightarrow \infty$, where y^n and x^n denote the sequences $y_1,\ldots,y_n$ and $x_1,\ldots,x_n$ of the observed data points y_t and x_t, respectively. It will certainly be useful to study additional questions like how fast can the convergence, required for consistency, take place in each class of models of interest;

ie, distributions $P(y|x)$, as well as the related but somewhat different question of how fast can 'learning' take place. Many such results are already available in the theory of estimation, see eg. Stone (1980), and they give guidance to what can be expected in a practical estimation job.

In this paper we first outline the MDL principle, which permits us to quantify the amount of learning from data with proposed models. For this, it is not necessary to assume that the data have been actually generated by any probabilistic or deterministic machinery at all, for we still can fit models and learn. This is in contrast with the previous formalizations where learning is defined only for data that are actually generated by a mathematically defined concept, which of course never is the case in reality. We illustrate the application of the MDL principle to a typical stochastic learning problem, where the features range over a continuum. This paradigm, which may be interpreted as the problem of learning the probability, as a function of the weight, with which a man thinks himself as being fat, includes a large number of analogous concrete instances. If we then assume for the purposes of analysis that the data have been obtained by sampling a certain sufficiently smooth distribution in a large family, we can give asymptotic bounds for learnability. By analogous and in fact simpler arguments one can obtain similar results for the cases where the feature space is finite or even countable.

2. The MDL Principle

In this section we summarize the relevant facts about code lengths, Shannon's theorem, and the MDL principle. Let A be a finite set of data sequences $X^n = \{x_1, \ldots, x_n\}$ of length n, where each data item x_i ranges over a finite set X. (The finiteness assumption can be relaxed.) A code is a one-to-one function $C : A \mapsto B^*$, where B is the binary 'alphabet' and B^* the set of all finite binary strings. There is an important practical restriction on the codes we are interested in, namely, that no codeword $C(a)$ is allowed to be a prefix of another $C(b)$. Such codes are called *prefix* codes. Moreover, we take the codes to be *complete*, which means that if we arrange the codewords in a binary tree structure, the set of the codewords $C(A)$ defines the set of leaves of a complete binary tree. The lengths of the codewords $L(a) \equiv |C(a)|$ of a complete prefix code satisfy the important Kraft-inequality

$$\sum_{a \in A} 2^{-L(a)} \leq 1 \tag{2.1}$$

with equality.

Suppose next that there is a probability distribution $P(a)$ defined on the set A. Shannon's fundamental coding theorem states (1) that for all prefix codes C the mean length is bounded from below by the entropy,

$$\sum_{a\in A} P(a)L(a) \geq -\sum_{a\in A} P(a)\log P(a) \equiv H(P), \tag{2.2}$$

and, moreover, (2) the equality is reached if and only if

$$L(a) = -\log P(a), \tag{2.3}$$

for all $a \in A$. Here, and in the rest of the paper, the logarithm is to the base two, and $0 \log 0 = 0$. This theorem settles quite definitely the problem of efficient communication of data through a noiseless channel, when the probability distribution for the data is known. The data to be communicated, if encoded as binary codewords, ought to have the code length satisfying (2.3). Although this cannot always be exactly done, because the codewords must have integer length, it can be achieved to a good approximation, and by (2.2) the optimal mean information rate, the entropy, will also then be nearly achieved.

The problem of estimation, and learning, too, as we shall argue, also deals with information and coding. However, it differs from the communication problem in that no distribution for the data is known. Rather, we are faced with a sort of inverse problem: Given the data we want to 'learn' a distribution which the data seem to obey. This by (2.3) appears to be equivalent to designing an 'optimal' prefix code for the data, except that since no unique distribution is given we cannot meaningfully talk about the mean code length. Fortunately, this is not too serious a difficulty if the set A is large, as it often is when it is taken as the set of sequences of some not too short a length, because then (2.2) will hold for the majority of sequences, generated by any distribution, even without the expectation operation. However, a search for an efficient code for the data without a distribution being given implies another more relevant change to the Shannon theory, namely, whatever code we, as an imagined encoder, design it must itself be somehow described to the decoder. This way, then, we may formulate a broad principle for estimation, the so-called principle of Minimum Description Length, or MDL. It calls for finding that model (= distribution) which permits the shortest encoding of the observed data together with the model itself. More precisely, if we consider a class of models $\mathcal{P} = \{P(x)\}$, we can write the code length obtainable with this class as

$$L(x^n|\mathcal{P}) = -\log \frac{\hat{P}(x^n)}{\sum_{z^n} \hat{P}(z^n)}, \tag{2.4}$$

where $\hat{P}(x^n)$ is the maximum value of $P(x^n)$ over $P \in \mathcal{P}$. Although there is a good estimate for the denominator for a large class of parametric families, Rissanen (1994), we use in this paper a coarser estimate of (2.4) (because the better formula was not known at the time this paper was first written),

$$L_{\mathcal{P}}(x^n) = -\log \hat{P}(x^n) + L(\hat{P}), \tag{2.5}$$

where $L(P)$ denotes the code length needed to encode the distribution P in the given class. Encoding the models is quite a different task from encoding data, since they lack natural distributions. Typically, the calculation involves encoding of integers in a prefix manner, which can be done in a near optimal manner by a 'universal' code length, Elias (1975), Rissanen (1983). With this optimally truncated real-valued parameters can also be encoded in an approximately optimal manner. For important model classes encoding of the real-valued parameters can be avoided altogether in the predictive form of coding as well as by the calculation of the 'stochastic complexity' (2.4), Rissanen (1994).

Notice that the final code length is still conditioned on the model class $\mathcal{P}$, which the same principle permits us to optimize as well - provided that we agree on some other ground language in which to encode the different model classes. Such a language could consist simply of a list of model classes, agreed upon between the encoder and the decoder. Each class might be defined by a set of free parameters of a fixed number, such as the set of all polynomials of degree $k = 0, 1, \ldots$. Or, more generally, we may permit the classes to be entirely different, the only requirement being that they can be described in a natural language with the usual mathematical notations. In such a case, unlike in the Bayesian philosophy, the 'prior', $2^{-L(\mathcal{P})}$, induced by the code length for the model class, is not for us to choose any way we wish, but it must necessarily reflect the complexity of the model class. Hence, the Bayesians' subjective 'prior' knowledge is replaced by 'common' knowledge, available both to the encoder and the decoder alike. The difference is quite important, for the MDL principle allows optimization of all the code lengths and hence the induced 'priors' while in the Bayesian philosophy they must be selected on other grounds, usually by mathematical tractability.

As a final remark, by assuming only a class of distributions as 'models' of data, known both to the encoder and the decoder, the question of optimal encoding calls for an extension of Shannon's coding theorem. Such an extension for parametric classes of distributions is given in Rissanen (1984) and (1986).

3. A Measure for Learning Rate

In our view learning is fundamentally the process of discovering constraints in the observed data. The constraints can be of many seemingly different kinds, but what we can precisely deal with are mathematical and probabilistic relations. In fact, families of distributions $\mathcal{P} = \{P(y^n, x^n)\}$, which may involve mathematical statements with parameters about the means and other statistical properties, cover just about all 'models' of interest, as we also call the distributions. As suggested by Yamanishi (1990), the relevant 'concept' of interest in learning theory appears to be the conditional distribution $P(y|x) = P(y,x)/P(x)$, induced by a joint distribution in the family, where $y \in Y$ and $x \in X$, the former set finite. It also appears to be adequate in many cases to assume independence of the feature occurrences $P(x^n) = \prod_{t=1}^n P(x_t)$ as well as the conditionals $P(y^n|x^n) = \prod_{t=1}^n P(y_t|x_t)$ so that also $P(y^n, x^n) = \prod_{t=1}^n P(y_t, x_t)$ holds. Such an assumption is well justified in the example we analyze in the following section, but for the theory itself no such restriction is needed. In the general case then $P(y^n, x^n) = \prod_{t=1}^n P(y_t, x_t|y^{t-1}, x^{t-1})$.

Constraints in the data have the immediate implication that they confine the data, or at least their bulk in a probabilistic sense, to a portion of the full range $Y^n \times X^n$. This, in turn, implies that, if we know the constraints, just as the case is in Shannon's theory, we can describe the data in a shorter way than what it takes to just write down each number, say, in a binary notation. The more restrictive the constraints are, the shorter the available code length is, and we can measure the amount of constraints by the decrease in the code length. This further suggests that, if we identify the amount of learning with the strength of the discovered constraints, we can measure both by the decrease in the code length; ie, with the unit of one bit. It is important to realize, however, that when calculating the decrease in the code length for the data obtainable with models in a proposed class, we must make sure that the code length refers to a code which has all the information needed for the correct decoding. One way to ensure this is to include in the total code length the length required to encode the model being used; ie, the constraints themselves, as indicated by (2.5).

In the case of interest in this paper the constraints we wish to learn are expressed by a class of models $P(y^n|x^n)$ restricting the data y^n, when the feature data x^n are given. The MDL principle of the type (2.5) calls for finding that distribution $\hat{P}(.|.)$, as a function of the data sequence (y^n, x^n), which minimizes the total code length

$$L(y^n|x^n, \mathcal{P}) = -\sum_{t=1}^{n} \log \hat{P}(y_t|x_t) + L(\hat{P}(.|.)), \tag{3.1}$$

where $L(\hat{P}(.|.))$ denotes the code length needed to encode the conditional distribution itself.

If we take (3.1) to represent a reasonable estimate of the shortest code length with which the data can be encoded, relative to the models in the considered class $\mathcal{P}$, we can measure the amount of learning per observation, the *learning rate*, either by the difference $\log|Y| - \frac{1}{n}L(y^n|x^n, \mathcal{P})$ or by

$$\frac{1}{n}L(y^n|x^n, \mathcal{P}) - \rho(P), \tag{3.2}$$

if the limit $\lim_{n\to\infty} \frac{1}{n}L(y^n|x^n, \mathcal{P}) = \rho(P)$ exists, where P is any distribution in $\mathcal{P}$ generating the data. Evidently, learning takes place as long as $\frac{1}{n}L(y^n|x^n, \mathcal{P})$ decreases. Hence, if the data are purely random so that no compression is possible, no learning takes place, which agrees with intuition: random strings have nothing to teach us. In the other extreme, algorithmically generated strings may be completely learnable in that we may discover the rule from a finite string. Strings in practice fall in between the two extremes, and learning never ends: The per symbol code length $\frac{1}{n}L(y^n|x^n, \mathcal{P})$ decreases forever with n to a non-zero limit. A fundamental question is to estimate the optimal asymptotic rate of learning for each considered model class, when we assume that the data are generated either by some distribution in the class or, more generally, by some distribution which is a suitable limit of the distributions in the class. For parametric classes of models with k free parameters we already know the answer, Rissanen (1984), (1986): The optimal asymptotic learning rate is $\frac{k}{2n}\log n$.

4. An Example

We consider an example where the data consist of pairs y_t, x_t for $t = 1, 2, \ldots, n$. The range of y_t is $Y = \{0, 1\}$ and that of x_t is a bounded interval of the real line. Such data could result from polling n men, in which case x_t is the weight of the tth man and y_t is 1, if he thinks himself to be fat; otherwise, $y_t = 0$. The objective is to learn a conditional binary probability distribution $P(y = 0|x)$ as a function of the weight x, representing the probabilistic concept of a man's perception of his obesity. Although one may argue that the t'th man's perception of his obesity might depend on how the preceding men regarded themselves, it seems reasonable to ignore this, which we do, and we adopt the independence assumption. For simplicity, we normalize the weights so that they all fall within the unit interval.

To formalize this learning problem we model the data sequence x_t as having resulted by sampling a histogram density function $f(x|\theta)$, obtained by subdividing the unit interval into m equal-length bins $\{I_i\}$. The density function is constant over each bin and assigns the probability P_i to the ith. Write $\theta = P_1, \ldots, P_m$ for the probability parameters which add up to unity. As to the conditional distributions, let p_{i0} be a parameter, $0 \leq p_{i0} \leq 1$, and take $P(y|x, p_{i0})$ for x in the ith bin as the Bernoulli distribution with the probability $P(0|x, p_{i0}) = p_{i0}$. For m a parameter, we define first the class of models $\mathcal{P}_m = \{P(y, x) = P(y|x, p_{i0})f(x|\theta) : x \in I_i, i = 1, \ldots, m\}$, and then $\mathcal{P} = \bigcup_m \mathcal{P}_m$. The intent then is to learn the conditional Bernoulli distribution $P(0|x)$ as a function of x, which we necessarily must approximate, in this case by a piecewise constant function $P(y|x, p_{i0})$, defined by the finite sequence $p_{10}, \ldots, p_{m0}$ of parameters for some value of m.

By the MDL principle the optimal piecewise constant conditional distribution is the one which permits the shortest code length (3.1) for the data $y^n|x^n$. An essential part in this is the choice of the number of bins m. With a finite number of observations available, it is not clear which value for m we ought to use, intuitively, because if m is too large, we do not have enough counts in the bins to obtain reliable estimates of the parameters p_{i0}, as it were, while if m is too small the estimated piecewise constant function will be too crude. First, by the assumption of independence the main question is how to encode the binary valued observations of y in each bin, for a given m. Let n_i denote the number of observations x_t that fall in the ith bin, and let n_{i0} be the number times y_t of these occurrences is 0. Since there are $\binom{n_i}{n_{i0}}$ such binary sequences, it is clear that the desired code length is a little more than the logarithm of this number. After all, although we know the counts n_i since we know x^n, we still need the counts n_{i0} in order to be able to decode each string. The encoding of the missing count n_{i0} for each i can be made with about

$\log(n_i + 1)$ bits, and we can encode the data y_t in the ith bin with length about

$$\log \binom{n_i}{n_{i0}} + \log(n_i + 1). \tag{4.1}$$

This is quite adequate except for short strings when the ratio n_{i0}/n_i is either close to 0 or 1. The more accurate code length (2.4) in this case is given by the formula, Rissanen (1994),

$$L(x^n) = nH(\{n_i/n\}) + \frac{1}{2} \ln \frac{n\pi}{2} + o(n),$$

where $H(p)$ is the binary entropy function.

Another useful way to obtain these code lengths is with predictive coding: Encode the first symbol with probability 1/2 or with one bit. Then, after having seen $n_{i0}(t)$ 0s of $n_i(t)$ symbol occurrences in the ith bin from the sequence y^t, x^t encode the next outcome y_{t+1} with probability

$$\hat{p}_{i0}(t) = (n_{i0}(t) + \alpha)/(n_i(t) + 2\alpha), \tag{4.2}$$

in case x_{t+1} falls in the *ith* bin and $y_{t+1} = 0$, or with the code length $-\log \hat{p}_{i0}(t)$, and with probability $\hat{p}_{i1}(t) = 1 - \hat{p}_{i0}(t)$ if $y_{t+1} = 1$. The sum of all these lengths for $\alpha = 1$ gives the formula (4.1) with the notations $n_i = n_i(n)$ and $n_{i0} = n_{i0}(n)$. With $\alpha = 1/2$ we get the more accurate formula, which we do not use in this paper for the reason given above. The term $\log(n_i + 1)$ in (4.1) represents an instance of the elusive but fundamental *model cost* associated with the ith bin, as seen in light of the observed data. It is an essential part in all estimation, and the lack of its formalization in traditional statistics has forced dealing with it only by ad hoc means.

We have a choice about how to proceed with the number of the bins: We can either let it be the same for all the observations, and optimize it, or we can let it be determined somehow as a function of y^t, x^t as t grows. In the former case we get the total code length using (4.1) as

$$L(y^n|x^n, \hat{m}) = \min_m \{\sum_{i=1}^{m} [\log \binom{n_i}{n_{i0}} + \log(n_i + 1)] + \log m + O(\log \log m)\}, \tag{4.3}$$

where the minimizing value for m is denoted by $\hat{m} = \hat{m}(y^n, x^n)$. The last two terms stem from encoding the number m with a prefix code. The estimated conditional distribution $P(y|x)$ is then given by the sequence (4.2) for $t = n$ and $\alpha = 1$. In the second case, for each sequence of the bin numbers the code length can be written as

$$L(y^n|x^n, m_0, \ldots, m_{n-1}) = -\sum_{t=0}^{n-1} \log \hat{P}(y_{t+1}|x_{t+1}, m_t), \tag{4.4}$$

where by (4.2)

$$\hat{P}(y|x_{t+1}, m_t) = \frac{n_{iy}(t)+1}{n_i(t)+2}, \tag{4.5}$$

and I_i denotes the ith of the m_t equal-length bins in which x_{t+1} falls. This time the optimum selection of the bin number m_t for the outcome y_{t+1} is not at all clear, for the selection must be done from the data y^t, x^t before y_{t+1} has been seen. In other words, the encoding is done predictively. As the term 'predictive' suggests, we must in effect guess what a good value for m_t ought to be, and perhaps as an aid use analysis. For instance, we can calculate the expected redundancy

$$R_m(y^t, x^t) = \sum_{i=1}^{m} \int_{I_i} f(x_{t+1}) D(P(0|x_{t+1}) \;||\; \hat{P}(0|x_{t+1}, m)) dx_{t+1}, \tag{4.6}$$

where for two Bernoulli distributions $p = Prob(0)$ and $\bar{p} = Prob(0)$,

$$D(p||\bar{p}) = p \log \frac{p}{\bar{p}} + (1-p) \log \frac{1-p}{1-\bar{p}}. \tag{4.7}$$

For a wide class of density functions f and conditional distributions $P(y|x)$ the redundancy turns out to depend asymptotically only on m and t, and hence we can minimize it over m with the result $\hat{m}_t = O(t^{1/3})$, which gives a sequence of tentative bin numbers. Moreover, asymptotically the resulting code length (4.4) turns out to be slightly shorter than (4.3). However, for real data, which have not been generated by any distribution, we cannot dismiss outright either the code length (4.3) or (4.4), with whatever sequence of bin numbers has been selected, for the result will necessarily depend on the particular observed data sequence.

In the rest of the paper we analyse the asymptotic behavior of both of the code lengths (4.3) and (4.4) by assuming that the data are generated by a joint distribution $P(y, x) = P(y|x)f(x)$, where $f(x)$ is a density on the unit interval such that $0 < c_0 \leq f(x) \leq c_1$, $c_0 < 1$, $c_1 > 1$, and having a bounded first derivative, say $|f'(x)| \leq c_f$. Assume further that $P(0|x)$ also satisfies $0 < c_2 \leq P(0|x) \leq c_3 < 1$ having a bounded first derivative, say $|P_x(0|x)| \leq c_P$. Let $\mathcal{M}$ denote the class of all such joint distributions. The following theorem, proved in Appendix A, gives then an asymptotically optimal number of bins in the order of magnitude as well as a bound for the asymptotic learning rate, in the case where only a single number of bins are used in the histogram density functions.

Theorem 1. For all joint distributions in $\mathcal{M}$, where $P(0|x)$ is not identically constant,

$$\hat{m} = \hat{m}(y^n, x^n) = O((n/\log n)^{1/3}) \quad a.s., \tag{i}$$

$$\frac{1}{n}L(y^n|x^n, \hat{m}) - \int H(P(0|x))f(x)dx = O((\frac{\log n}{n})^{2/3}) \quad a.s., \tag{ii}$$

where $H(p) = -p\log p - (1-p)\log(1-p)$, and

$$\int f(x)dx \sum_{y=0,1} P(y|x)\log\frac{P(y|x)}{\hat{P}(y|x)} = O((\frac{\log n}{n})^{2/3}). \tag{iii}$$

Remarks. The second term in (ii) is achieved when we have learned the target conditional distribution $P(y|x)$ completely. If this is identically constant there is obviously nothing to be gained by increasing the number of bins; hence the exception in the theorem. The right hand side, then, gives in the order of magnitude the learning rate, which in view of (iii) also gives the rate with which the estimated conditional distribution approaches the target, when the distance is calculated in terms of the Kullback-Leibler distance function. Hence we see that our measure for the amount of learning admits a second intuitively appealing interpretation.

The analysis of the second case is more complicated and we prove in Appendix B the following theorem, which gives the asymptotic learning rate in expectation, rather than in almost sure sense.

Theorem 2. Let $m_t = \lceil t^{1/3}$. Then for all distributions in $\mathcal{M}$ the expected redundancy in (4.6) satisfies

$$\frac{1}{n}\sum_{t=1}^{n} ER_{m_t}(y^t, x^t) \leq O(n^{-2/3}).$$

Remark. It is cumbersome but possible to prove that the right hand side bound cannot be improved upon, except for rare data generating densities f and functions $P(0|x)$, so that it indeed gives the asymptotic learning rate.

Acknowledgement. Bin Yu gratefully acknowledges the partial support from Army Research Office Grant DAAL03-91-G-007.

Appendix A. Proof of Theorem 1. If we write $L(y^n|x^n, m)$ for the code length obtainable with m bins, we get from (4.3) with Stirling's formula

$$\frac{1}{n}L(y^n|x^n, m) = \sum_{i=1}^{m}[\frac{n_i}{n}H(\frac{n_{i0}}{n_i}) + \frac{1}{2}\frac{\log(n_i+1)}{n}] + \frac{\log m}{n} + O(\frac{\log\log m}{n}), \qquad (A.1)$$

which holds in probability as well as almost surely, Rissanen et al. (1989). There are two types of convergence to consider. The first and the simpler one concerns the average

$$P_{i0} = \frac{\int_{I_i} P(0|x)f(x)dx}{P_i}, \qquad (A.2)$$

where I_i denotes the ith bin of width $1/m$ and $P_i = \int_{I_i} f(x)dx$. We wish to show that for each x

$$P_{i(x)0} \to P(0|x),$$

as $m \to \infty$, where $i(x)$ is the bin of width $1/m$ which includes x. Let for x in the ith bin, $\epsilon(x) = P_{i0} - P(0|x)$. Then by the mean value theorem

$$\epsilon(x) = P_x(0|\xi)(x - a_i)$$

where a_i is a point within the bin such that $P_{i0} = P(0|a_i)$, and ξ is another point that depends on x and a_i. Further, with the notation (4.7) and Taylor's series expansion we get

$$D(P(0|x)||P_{i0}) = \frac{1}{2}D_{xx}(\eta)\epsilon^2(x) = G_{ix}(x - a_i)^2, \qquad (A.3)$$

where the double derivative is evaluated at a point η within the ith bin that also depends on a_i and x, and $G_{ix} = \frac{1}{2}D_{xx}(\eta)P_x^2(0|\xi)$. Clearly, $0 \leq G_{ix} \leq \alpha$ for some constant α. By multiplying both sides of (A.3) by $f(x)$ and integrating over the unit interval we get

$$\sum_i P_i H(P_{i0}) - \int H(P(0|x)f(x)dx = \sum_i \int_{I_i} f(x)G_{ix}(x - a_i)^2 dx. \qquad (A.4)$$

Unless $P(0|x)$ is identically constant there is an interval where its derivative does not vanish. For large enough m such an interval will span across the order of m bins, for each of which the integral in the right hand side is of the order $1/m^3$. Hence, the right hand side is of the order $1/m^2$. We can then write from (A.1) and (A.4)

$$\begin{aligned}\frac{1}{n}L(y^n|x^n,m) - \int H(P(0|x))f(x)dx = \sum_{i=1}^{m}[\frac{n_i}{n}H(\frac{n_{i0}}{n_i}) - P_i H(P_{i0})] \\ + \frac{m\log\frac{n}{m}}{2n} + \frac{\log m}{n} + O(1/m^2),\end{aligned} \tag{A.5}$$

almost surely.

The second type of convergence we need to consider is that of the count ratios

$$\begin{aligned}\frac{n_{i0}}{n_i} &\to P_{i0} \quad a.s. \\ \frac{n_i}{n} &\to P_i \quad a.s.\end{aligned} \tag{A.6}$$

Since n_i has the binomial distribution $Bin(n, P_i)$ the probabilistic convergence rate for the second ratio is given by Bennett's inequality, Pollard (1984), as

$$P\{|\frac{n_i}{n} - P_i| \geq c\} \leq 2e^{-Bnc^2/m}, \tag{A.7}$$

where B is a constant determined by the lower bound c_0 for the density functions. To get an analogous bound for the first ratio in (A.6) is a bit more complicated, because n_i itself is a random variable. However, we have

$$P\{|\frac{n_{i0}}{n_i} - P_{i0}| \geq c\} = \sum_{k=0}^{n} P\{|\frac{n_{i0}}{k} - P_{i0}| \geq c|n_i = k\}P\{n_i = k\},$$

where the first factor is taken as unity for $k = 0$. Conditioned on $n_i = k$ the count n_{i0} has the binomial distribution $Bin(k, P_{i0})$, so that it satisfies Bennett's inequality with the constant $B = B_1$ now depending on the lower bound on the numbers P_{i0}. Hence,

$$\begin{aligned}P\{|\frac{n_{i0}}{n_i} - P_{i0}| \geq c\} &\leq 2\sum_{k=0}^{n} e^{-B_1kc^2/m}P\{n_i = k\} \\ &< 2P(\{|n_i - np_i| \geq np_i/2\} + 2\sum_{|k-np_i|<np_i/2} e^{B_1kc^2/m}P\{n_i = k\} \\ &\leq 4e^{-Bnp_i^2/4m} + 2e^{-B_1np_ic^2/2m} \\ &\leq 4e^{-Bc_0^2n/4m} + 2e^{-B_1c_0c^2n/2m}\end{aligned} \tag{A.8}$$

It follows from (A.7) and (A.8) that the first term in the right of (A.5), written as $T_n(m)$, satisfies

$$P\{||T_n(m)| \geq m\epsilon|\} \leq me^{-K\epsilon^2 n/m}, \tag{A.9}$$

where K is a constant. Hence for all nonconstant $P(0|x)$ and all strings, except for those in a set whose probability is bounded by this exponential bound,

$$\frac{1}{n}L(y^n|x^n, m) - \int H(P(0|x))f(x)dx = +\frac{m\log\frac{n}{m}}{2n} + \frac{\log m}{n} + O(1/m^2), \tag{A.10}$$

which is minimized asymptotically for m given by $\hat{m}_n = O((\frac{n}{\log n})^{1/3})$. Moreover, for this value

$$\sum_{n\geq 1} P\{||T_n(\hat{m}_n)| \geq \epsilon|\} < \infty$$

for every ϵ, which implies the almost sure convergence claims in the theorem.

Appendix B. Proof of Theorem 2. With the notations of Appendix A the redundancy (4.6) can be written as

$$R_m(y^t, x^t) = \sum_{i=1}^{m}\int_{I_i} f(x)D(P(0|x)\;||\;P_{i0})dx + \sum_{i=1}^{m}\int_{I_i} f(x)\sum_y P(y|x)\log\frac{P_{iy}}{\hat{P}_{iy}}dx, \tag{B.1}$$

where we also wrote

$$\hat{P}_{iy} = \frac{n_{iy}(t)+1}{n_i(t)+2}.$$

We showed in Appendix A that the first term is of the order $O(m^{-2})$. We denote the second term as $V_t(m) = V_{t1}(m) + V_{t0}(m)$, the two parts corresponding to the values $y = 1$ and $y = 0$, respectively. Hence,

$$V_{t1}(m) = \sum_{i=1}^{m} P_i P_{i1}\log\frac{P_{i1}}{\hat{P}_{i1}}. \tag{B.2}$$

By Taylor's expansion

$$\log\frac{\hat{P}_{i1}}{P_{i1}} = \frac{\hat{P}_{i1} - P_{i1}}{P_{i1}} - \frac{1}{2}\frac{(\hat{P}_{i1} - P_{i1})^2}{P_{i1}^2} + \ldots = T_1 + T_2 + \ldots.$$

To evaluate the expected values of the two terms we need the Lemma:

Lemma. For X having the binomial distribution $Bin(n,p)$,

$$E\frac{1}{X+2} = \frac{1}{(n+1)p}(1-(1-p)^{n+1}) = \frac{1}{np} + O(n^{-2}) \qquad (i)$$

$$E\frac{1}{(X+2)^2} = \frac{1}{n^2p^2} + O(n^{-3}) \qquad (ii)$$

The first part (i) follows by direct evaluation and using the fact that

$$\binom{n}{k}\frac{1}{k+1} = \binom{n+1}{k+1}\frac{1}{n+1}.$$

The second part follows by applying the same fact together with (i) and noting that

$$(1-\frac{1}{k+1})\frac{1}{(k+1)(k+2)} \leq \frac{1}{(k+2)^2} \leq \frac{1}{(k+1)(k+2)}.$$

By taking first the conditional expectation with respect to n_{i1}, given n_i, and then the expectation with respect to n_i, we get with (i) of the Lemma

$$ET_1 = \frac{1-2P_{i1}}{tP_iP_{i1}} + O(t^{-2}).$$

Similarly with (ii) of the Lemma

$$ET_2 = -\frac{1}{2}\frac{1-P_{i1}}{tP_iP_{i1}} - \frac{1-2P_{i1}}{tP_iP_{i1}^2} + O(t^{-2}).$$

These will give us $EV_{t1}(m)$ and similarly $EV_{t0}(m)$ with their sum as

$$EV_t(m) = \frac{m}{2t} + \frac{1}{t}\sum_{i=1}^{m}\frac{(P_{i0}-P_{i1})^2}{P_{i0}P_{i1}} + O(m/t^2) \leq C\frac{m}{t} + O(m/t^2). \qquad (B.3)$$

for a constant C, $C > 1$, that depends on the bounds given in the model class. Adding the size of the first term in (B.1), $O(m^{-2})$, to the upper bound of second term given in (B.3), and minimizing with respect to m, we get the minimizing number of bins as $O(t^{1/3})$. The minimized mean redundancy is then $O(t^{-2/3})$. Summing this from $t = 1$ to n and dividing the result by n gives the inequality claimed in the theorem.

References

Angluin, D., Laird, P. (1988), 'Learning from Noisy Examples', Machine Learning 2, 343-370

Elias, P. (1975), 'Universal Codeword Sets and Representations of the Integers', IEEE Trans. Inf. Theory, Vol **IT-21**, no. 2, 194-203

Haussler, D. (1989), 'Generalizing the PAC Model for Neural Net and Other Learning Applications', Technical Report UCSC CRL-89-30, University of Santa Cruz, September 1989

Kearns M., Li M. (1988), 'Learning in the Presence of Malicious Errors', Proc. of the 20th Annual ACM Symposium on Theory of Computing', Illinois, May 1988

Pollard, D. (1984), *Convergence of Stochastic Processes*, Springer-Verlag, New York, (215 pages)

Rissanen, J. (1983), 'A Universal Prior for Integers and Estimation by Minimum Description Length', Annals of Statistics, Vol. **11**, No. 2, 416-431

Rissanen, J. (1984), 'Universal Coding, Information, Prediction, and Estimation', IEEE Trans. Inf. Theory, Vol. **IT-30**, No. 4, 629-636

Rissanen, J. (1986), 'Stochastic Complexity and Modeling', Annals of Statistics, Vol **14**, 1080-1100

Rissanen, J. (1989), *Stochastic Complexity in Statistical Inquiry*, World Scientific Publ. Co., New Jersey, (175 pages)

Rissanen, J. (1994), 'Stochastic Complexity and Fisher Information' (submitted to IEEE Trans. Inf. Theory)

Rissanen, J., Speed, T., Yu, B. (1989), 'Density Estimation by Stochastic Complexity', IEEE Trans. Inf. Theory, Vol. **IT-38**, No. 2, 315-323

Rivest, R.L. (1987), 'Learning Decision Lists', Machine Learning, 2, 229-246

Stone, C.J. (1980), 'Optimal Rates of Convergence for Nonparametric Estimators', Annals of Statistics, Vol. 8, No. 6, 1348-1360

Suzuki, J. (1990), 'Generalization of the Learning Method for Classifying Rules with Consistency Irrespective of the Representation Form and the Number of Classified Patterns', 1990 Int. Symposium on Information Theory and its Applications, Hawaii, November 1990

Valiant, L.G. (1984), 'A Theory of the Learnable', Comm. of the ACM, 27, 1134-1142

Yamanishi, K. (1990), 'A Learning Criterion for Stochastic Rules', Proc. of the Third Annual Workshop on Computational Learning Theory, August 1990

Pac Learning, Noise, and Geometry

Robert H. Sloan*
Department of EE & Computer Science
851 S. Morgan St. Room 1120
University of Illinois at Chicago
Chicago, IL 60607-7053 USA

Abstract

This paper describes the probably approximately correct model of concept learning, paying special attention to the case where instances are points in Euclidean n-space. The problem of learning from noisy training data is also studied.

*Supported by NSF grant CCR-9108753. Email: `sloan@eecs.uic.edu`

1 Introduction

We will be concerned with that part of Machine Learning referred to as *concept learning* or *learning from examples*. The basic paradigm is as follows.

1.1 Example: An afternoon stroll

It is a bright, sunny afternoon. I take my robot out for a walk with me. We stop and sit down on a bench in a busy part of town. For every person who walks by us I say to my robot either, "That's a male," or "That's not a male," as the case may be. We stay quite some time, and eventually go home.

The following afternoon the weather is the same, and we go back to the same bench. This time, for every person who walks by, my robot says to me either "That's a male," or "That's not a male." My robot is correct ninety-six percent of the time.

1.2 The basic problem

The robot in the above fable (description of future research plans?) has come up with some rule for splitting the entire universe under consideration into two classes. That is a particular case of the following general class of problems.

There is some universe of objects under consideration, called the *instance space*. Frequently studied instance spaces include Euclidean n-space, bit vectors of some fixed length, and the integers. Of course, the set of all people is also a perfectly good instance space.

A *concept* is a subset of the instance space. An *example* of concept c consists of an instance, x, together with a label $c(x)$ that is 1 if $x \in c$ and 0 otherwise. (Throughout we will refer to concepts interchangeably as sets and $\{0, 1\}$-valued functions.) If $c(x) = 1$ then x is a *positive instance* of concept c; otherwise x is a *negative instance*. The particular concept we are trying to learn is usually called the *target concept*. We always assume that the learner knows a priori that the target concept comes from particular *concept class*, or set of concepts. This is not a restriction, since the concept class could simply be the power set of the instance space.

Our learner is given a set of examples, or *sample* of the target concept, and asked to output the target concept (or some good approximation to it). In the example above, people constitute the instance space,[1] and the target concept is the set of male people. The robot demonstrates that it has found a good approximation to the target concept by providing a fairly accurate classification of previously unseen instances (people).

For the problems considered in this paper, that ability to classify is the point of the learning process. The examples given to the learner should be thought of as a training phase, so that in a future performance phase the learner can accurately classify instances for itself.

This problem has been studied by the Artificial Intelligence (AI) community since the very beginning of that field—primarily in an empirical fashion. More recently, the theoretical computer science community has taken an interest in this field. The work by this community began with Gold [11] back in the 1960s. (See Angluin and Smith [3] for an excellent survey of work through the early 1980s.) A seminal paper by Valiant [26] in 1984 prompted intensified effort. The goal of the work by the theory community, not surprisingly, is to provide precise models of the problem, and good algorithms to solve the problems so modeled (or proofs of the hardness of those problems).

The rest of this paper is organized as follows. The next section explains the method that theoretical computer scientists use for measuring problem difficulty. Then we will give a detailed definition of Valiant's model of concept learning (pac learning), and discuss why it is an attractive model. The remainder of the paper will concentrate on algorithms for that model, beginning briefly with finite instance spaces, and then concentrating on

Geometry the case where the instance space is Euclidean n-space, and the target concept is some region, and

Noise what happens to these algorithms when the examples used in the training phase are corrupted by noise.

---

[1] Perhaps it would be more accurate to say that the instance space consisted of the cross product of some set of discrete and continuous attributes (including, say, height, width, clothing type, etc.) that the robot actually observed.

2 Asymptotics

The particular interest of theoretical computer scientists studying concept learning, as opposed to say, statisticians, is finding *computationally efficient* learning algorithms. We will be concerned with two different measurements of efficiency:

1. The number of training examples needed.
2. The amount of computing time needed.

The standard definition of "efficient" consumption of whatever resource is under consideration is some amount polynomial in the "size of the problem."

For that to have any meaning, however, we must discuss, of course, the definition of problem size, but also the asymptotic difficulty of our problems. If we just consider any one fixed problem, it always has a "constant time" solution. For instance, the infamous Traveling Salesman problem for a fixed graph of 1000 nodes can be solved in constant time. (Of course, the "constant" in that case might be something around 2^{1000}.)

Thus instead of examining algorithms for one fixed learning problem, we always examine algorithms for an infinite sequence of learning problems. There are two different ways we can obtain such a sequence—by allowing the size/dimension of the instance space to grow, or by allowing the complexity of the concepts we consider to grow. We will now describe both these approaches briefly, but will concentrate on the first approach for the remainder of this paper.

2.1 Efficiency with respect to instance space dimension

Using this method, we parameterize the instance space by some natural dimension. Thus our instance space $X = \cup_{n=1}^{\infty} X_n$. For the sake of concreteness, in the rest of this paper we will always assume that our instance space X_n is one of E^n (Euclidean n-space), $[0,1]^n$, or $\{0,1\}^n$. We set $\mathbf{C} = \{(X_n, \mathcal{C}_n)\}_{n\geq 1}$, where $\mathcal{C}_n \subseteq 2^{X_n}$. We will use the term concept class to refer both to $\mathbf{C}$ and a particular $\mathcal{C}_n$. Note for future reference that when the literature refers to finite concept classes (or instance spaces), it means concept classes where $|\mathcal{C}_n|$ (or $|X_n|$) is finite for each n.

We are interested only in computationally efficient learning. Therefore, in addition to other assumptions, we always assume that our concept class has the property that there is a polynomial (in n) time algorithm for determining whether $x \in c$, for any $x \in X_n$ and $c \in \mathcal{C}_n$.

2.2 Efficiency with respect to concept complexity

The other way to make our learning problems asymptotic is to take a fixed instance space, but to allow concepts to grow arbitrarily complex. For instance, we could fix the instance space to be the real plane, and let the entire concept class be the set of all convex polygons. We could then make $\mathcal{C}_n$ the set of convex n-gons, or alternatively, the set of convex polygons with at most n edges. This approach to pac learning is explored by Linial, Mansour and Rivest, and by Blumer et al. [8, 19].

3 Pac learning

In this section we define Valiant's model of concept learning [26]. Angluin dubbed this model "probably approximately correct learning," [1] and it is thus referred to as *pac learning.*

In short, an algorithm pac learns from examples if it can, in a feasible amount of time, find (with high probability), a rule that is highly accurate. Now we must define what we mean by such terms as "find a rule," "with high probability," and "highly accurate."

For the purposes of this discussion, fix a concept class $\mathbf{C} = \{(X_n, \mathcal{C}_n)\}_{n \geq 1}$. First we define formally what it means for one concept in $\mathcal{C}_n$ to be an accurate approximation of the target concept (or any other concept, for that matter), and then we define pac learning itself.

DEFINITION. Fix n, and a probability distribution $\mathbf{P}$ on X. For any $0 \leq \epsilon \leq 1$, we say that concept c_1 is an *ϵ-approximation* of concept c_0 if and only if

$$\sum_{c_1(x) \neq c_0(x)} \mathbf{P}(x) \leq \epsilon. \tag{1}$$

The goal of our learning algorithm will be to find an ϵ-approximation of the target concept.

We assume that our learning algorithm has available to it a black box called EXAMPLES, and that each call to the black box returns a labeled example, (x, s), where $x \in X_n$ is an instance, and s is either 1 or 0 according to whether x is a positive or negative instance of the target concept. Furthermore, the EXAMPLES box generates the instances x according to some fixed probability distribution $\mathbf{P}$ on X_n. We make no assumptions whatsoever about the nature of $\mathbf{P}$, and our learner is not told what $\mathbf{P}$ is.

DEFINITION. Let $\mathbf{C} = \{(X_n, \mathcal{C}_n)\}_{n \geq 1}$ be a concept class. Algorithm A *probably approximately correctly learns (pac learns)* $\mathbf{C}$ if and only if for every $n \geq 1$, for every $c \in \mathcal{C}_n$, for every probability distribution on X_n, for every $0 < \epsilon, \delta \leq 1$, Algorithm A, given only n, ϵ, δ and access to EXAMPLES(c), meets the following two criteria.

Learning criterion Algorithm A outputs some representation of a concept c' such that

$$\Pr\left[c' \text{ is an } \epsilon\text{-approximation of } c\right] \geq 1 - \delta \tag{2}$$

where the probability is taken over the output of EXAMPLES and any coin tosses A may make.

Efficiency criterion The running time of A is bounded by some polynomial function of $1/\epsilon$, $1/\delta$, and n.

We say that a concept class $\mathbf{C}$ is *pac learnable* if there exists some algorithm that pac learns $\mathbf{C}$.

We note here for later use the following weaker definition.

DEFINITION. *Statistical pac learning or s-pac learning* has the same definition as pac learning except that the efficiency criterion requires the sample complexity, rather than the running time, to be bounded by by some polynomial function of $1/\epsilon$, $1/\delta$, and n.[2]

4 Discussion of the definition

Intuitively, we are saying that the learner is supposed to do the following:

[2]Terminology in this field is not yet completely settled. Many authors use the term pac learning for what we are calling statistical pac learning, and polynomial pac learning for what we are calling pac learning.

1. Ask nature for a random set of examples of the target concept.

2. Run in polynomial time.

3. Output some concept that with high probability agrees with the target concept on most of the instances.

We think of Nature as providing examples to the learner according to the (unknown) probability with which the examples occur in Nature. Though the learner does not know this probability distribution, he does know that the concept he outputs needs to closely approximate the target concept for only this probability distribution.

One of the great strengths of pac model is that it places no restrictions whatsoever on nature of the probability distribution on the instance space. For this reason, the other common name for this model besides pac learning is *the distribution-free model* of (concept) learning.

The two probabilistic quantifiers in the definition are necessary to make the problem possible to solve. Intuitively, there may be some extremely bizarre but low probability examples that occur in Nature, and it would be unreasonable to demand that the learner's output concept classify them correctly. Hence we require only approximate correctness. Moreover, since the examples the learner receives from Nature are drawn stochastically, there is some small but nonzero chance that those examples will be wildly unrepresentative. Therefore we cannot demand that the learner *always* output an approximately correct concept; we require that the learner do so only with high probability.

The notion of statistical pac learning was studied in the statistical pattern recognition literature before the notion of pac learning was introduced in 1984. The difference between the two is that statistical pac learning requires us to be efficient in our consumption of examples but not in our computation time. Notice that any algorithm that pac learns must statistically pac learn since the sample complexity of an algorithm is a lower bound on its running time. Thus the strongest infeasibility results we can obtain are showing the infeasibility of statistical pac learning. In general, it is often easier to prove theorems about statistical pac learning than about pac learning.

The excellent survey article of Kearns et al. gives a more lengthy discussion of the entire pac learning model, and also provides an overview of early results obtained using this model [18]. Blumer et al. and Valiant both contain additional material on the motivation for the pac learning model [8, 26].

5 Variations on the definition of pac learning

The pac learning model can be altered in many ways. For instance, instead of having a single probability distribution on the whole instance space and a single box EXAMPLES, there can be two separate probability distributions, one each for positive examples and negative examples, and two boxes, EXAMPLES$^+$ and EXAMPLES$^-$, each giving only one kind of instance. In this case, the learning algorithm is allowed to make calls to either box, and its hypothesis is required to come within ϵ on both positive and negative examples. This model is called the two-button model; the model we are using is called the one-button model.

There are in fact numerous different minor variations on pac learnability. Haussler et al. compare many of these variations, and show them to be substantially equivalent [13].

6 An example of pac learning

We now exhibit an algorithm that pac learns monomials [26]. The instance space X_n is $\{0,1\}^n$ for some positive n. The concept class is the set of all boolean expressions in n variables that can be represented as monomials (simple conjunctions of literals).

Algorithm A for pac learning is as follows: Algorithm A initializes its hypothesis to be the conjunction of all $2n$ literals, $x_1\bar{x}_1 \cdots x_n\bar{x}_n$. Algorithm A then makes m calls to EXAMPLES, where

$$m = \frac{1}{\epsilon}\left(n \ln 3 + \ln(1/\delta)\right).$$

For each positive example, $(instance, 1)$, Algorithm A crosses off x_i from its hypothesis if x_i is 1 in $instance$, and crosses off $\bar{x}_i$ if x_i is 0 in $instance$. Algorithm A simply ignores negative instances. The output of A is the monomial consisting of whatever literals have not been crossed off after the m

instances are seen. (If all of the m instances are negative, then A's hypothesis is $x_1\bar{x}_1 \cdots x_n\bar{x}_n$, which is always false.)

The running time of of A is clearly polynomial in $1/\epsilon, 1/\delta$, and n, so A meets the efficiency criterion for pac learning. In Section 7 we will give a general result that proves that Algorithm A meets the learning criterion for pac learning. (In fact, a somewhat better bound for m can be obtained by making a direct combinatorial argument.)

Note for later reference that the correctness of Algorithm A is critically dependent on the data from EXAMPLES being noise free. One negative instance, mistakenly labeled positive, would most likely cause the algorithm to output some totally bogus hypothesis.

7 Pac learning with finite concept classes

Let us briefly examine sufficient conditions for a concept class to be pac learnable. A starting place place is any concept class of very small cardinality. In the extreme, consider the case of a concept class containing only two concepts, c_1 and c_2.

Recall that we always assume it is computationally efficient to determine whether any particular instance is contained in a given concept. Now all we need to do is draw a sufficiently large sample from EXAMPLES, and select whichever concept is consistent with the sample, or, if both concepts are consistent, choose either one arbitrarily.

We can extend this line of reasoning to considerably larger (finite) concept classes. Our treatment follows Blumer et al. [7]. Consider an algorithm that draws a sample of size $m(|\mathcal{C}|, \epsilon, \delta)$, and then outputs any concept $c \in \mathcal{C}$ that is consistent with the sample. Ignoring for a moment the issue of computation time, what is a lower bound on the function m that insures that the learning algorithm meets the statistical pac learning condition?

DEFINITION. Fix a target concept and a probability distribution on the instance space. We say that a concept is ϵ-*good* if it is an ϵ-approximation of the target concept, and ϵ-*bad* if it is not.

We want to guarantee that the probability that the algorithm returns an ϵ-bad concept is at most δ. The probability that a particular ϵ-bad concept would be consistent with m randomly drawn examples is bounded above by

$$(1-\epsilon)^m.$$

A crude upper bound on the probability that *any* ϵ-bad concept is consistent with m randomly drawn examples is

$$|\mathcal{C}|\,(1-\epsilon)^m.$$

Hence we want to choose m large enough to insure that

$$|\mathcal{C}|\,(1-\epsilon)^m \le \delta. \tag{3}$$

Taking logs of both sides, and solving for m yields

$$m \ge \frac{1}{-\ln(1-\epsilon)}\left(\ln(|\mathcal{C}|) + \ln(1/\delta)\right). \tag{4}$$

We can write a simpler inequality that implies inequality (4) by noticing that by Taylor's Theorem

$$\begin{aligned}\ln\left(\frac{1}{1-\epsilon}\right) &> \ln 1 + \epsilon\frac{1}{(1-\epsilon)} + \frac{\epsilon^2}{2!}\frac{1}{(1-\epsilon)^2}\\ &> \frac{\epsilon}{1-\epsilon}\\ &> \epsilon.\end{aligned}$$

Hence, $1/\epsilon > 1/-\ln(1-\epsilon)$, and we have shown that

Theorem 1 *Fix a concept class* $\mathbf{C} = \{(X_n, \mathcal{C}_n)\}_{n\ge 1}$. *Any algorithm which on input* n, ϵ, *and* δ *makes*

$$m = \frac{1}{\epsilon}\left(\ln(|\mathcal{C}_n|) + \ln\left(\frac{1}{\delta}\right)\right). \tag{5}$$

calls to EXAMPLES *and returns any concept consistent with that sample statistically pac learns* $\mathbf{C}$.

We say that m is an upper bound on the *sample complexity* of the concept class $\mathbf{C}$. More generally we say that an arbitrary concept class $\mathbf{C}$ has sample complexity at most m if there is a learning algorithm with sample complexity m that statistically pac learns $\mathbf{C}$.

As long as $|\mathcal{C}_n| = O\left(2^{p(n)}\right)$, for some polynomial p, then $\mathbf{C}$ has polynomial sample complexity, and is statistically pac learnable. (If $|\mathcal{C}_n| = O\left(q(n)\right)$, for some polynomial q, then $\mathbf{C}$ is trivially pac learnable.) If $\mathbf{C}$ has polynomial sample complexity, and if the problem of finding a concept from **C** consistent with a given sample can be solved in polynomial time, then **C** is pac learnable.

There are 3^n different monomials on n boolean variables, so the concept class of monomials clearly has polynomial sample complexity. On the other hand, we obviously cannot exhaustively examine all 3^n monomials in polynomial time. The algorithm for learning monomials given above in Section 6 overcomes this problem by finding a consistent concept in a more clever manner.

8 Remark on the form of the sample complexity

The general form of Equation (5),

$$\frac{1}{\epsilon}\ln\left(\frac{1}{\delta}\right) + \frac{1}{\epsilon}\,(\text{term without } \delta \text{ concerning concept class}),$$

is one that crops up time and again in formulas for sample complexity. In this section we will attempt to see why this is so, and especially why the dependence on δ is almost always of the form $(1/\epsilon)\ln(1/\delta)$.

The explanation begins with the fact that Equation (5) is indeed a choice that makes the inequality

$$(1-\epsilon)^m \leq \delta \tag{6}$$

true. It is easy to see that Inequality (6) must always be true. For any concept class $\mathbf{C} = \{(X_n, \mathcal{C}_n)\}_{n\geq 1}$, we can always add some new distinguished element x to every instance space, and for every concept c in the class add a new concept $c \cup \{x\}$. If probability weight ϵ is placed on x, then our sample must always contain x, because the output concept must classify x correctly to meet the learning criterion, and there is no way to know anything about the classification of x without actually seeing its label. Hence we have a clear lower bound on sample complexity of $(1/\epsilon)\ln(1/\delta)$.

Can we push this line of reasoning farther, and get a larger lower bound than $(1/\epsilon)\ln(1/\delta)$? The most extreme form of this idea would be to design an instance space of size $n = 1/\epsilon$, put probability weight ϵ on each instance, and make the concept class the full power set of the instance space. Now we must draw enough examples to insure that we see all n instances at least once with probability at least $1-\delta$. This is the well known coupon collector's problem.[3] It turns out that we have the convergence

$$\Pr\left[cn + n\ln n \text{ draws are sufficient}\right] \to e^{-e^{-c}}. \tag{7}$$

(In fact, the convergence is quite rapid.)

Choosing $c = \ln(1/\delta)$ makes the right-hand side of Equation (7) equal to $e^{-\delta}$, which is approximately $1-\delta$, since for a of small absolute value we have $e^a \approx 1 + a$. Thus even in this case we still need to draw only

$$\frac{1}{\epsilon}\ln\left(\frac{1}{\delta}\right) + \frac{1}{\epsilon}\ln\left(\frac{1}{\epsilon}\right)$$

examples to have a $1-\delta$ probability of success.

9 Continuous instance spaces: the VC dimension

Consider now a typical geometric learning problem, $\mathbf{C} = (E^n, B^n)$, where B^n represents the set of all axis-parallel boxes contained in E^n. Now our concept class has uncountable cardinality, so the method of Theorem 1 cannot work, nor can any method based on the size of the concept class.

It turns out, however, that a combinatorial parameter of set systems can take the place of the cardinality of the concept class. In fact, this parameter, the Vapnik-Chervonenkis dimension [29], hereinafter VC dimension (or in symbols VCD$(\cdot)$), provides both a necessary and sufficient condition for the existence of an algorithm that can statistically pac learn just by drawing a fixed sample size. Since the VC dimension is a combinatorial parameter of set systems, we define it in terms of sets, though of course the sets we are interested are concepts.

9.1 VC dimension definition

DEFINITION. Fix a domain X, and let $\mathcal{C} \subseteq 2^X$. A set $S \subset X$ is *shattered* (by $\mathcal{C}$) if for each subset $S' \subseteq S$, there is a set $c \in \mathcal{C}$ which contains all of S', but none of the points in $S - S'$.

[3]So called because another way of stating it is, "A cereal manufacturer is putting n distinct coupons into cereal boxes with equal probability. How many boxes of cereal must one buy to insure getting at least one of each coupon?"

Remark: The term shatter is now well established. However, as Pollard points out [22], the right picture to keep in mind is not really S being broken into lots of tiny pieces by $\mathcal{C}$. Rather, one should imagine a diligent $\mathcal{C}$ picking out all the different subsets of S.

DEFINITION. The *VC dimension* of $\mathcal{C} \subseteq 2^X$ is the cardinality of the largest set of points from X shattered by $\mathcal{C}$.

Examples: The class of all rectangular regions in the plane has VC dimension 4. The class B^n on E^n has VC dimension $2n$. The class of all spheres in E^n has dimension $n+1$. For any finite concept class $\mathcal{C}$, we have $VCdim(\mathcal{C}) \leq \log(|\mathcal{C}|)$.[§] There are now a number of papers which calculate the VC dimension of different classes [4, 9, 14, 16].

The VC dimension was first studied in connection with statistical pattern recognition. Pollard and Vapnik have both written good books discussing it from that point of view [22, 28]. The first source that I am aware of to point out that it has some connection to efficient concept learning is Pearl [20].

9.2 Algorithms

For our purposes, the key fact about the VC dimension is that a fixed concept class has polynomial sample complexity if and only if it has finite VC dimension, or a sequence of concept classes $\mathcal{C}_n$ has polynomial sample complexity if and only if $\text{VCD}(\mathcal{C}_n)$ is polynomial in n. In particular, Blumer et al. showed [8]

Theorem 2 *For any instance space X and "well-behaved" $C \subseteq 2^X$, any learning algorithm which obtains a sample of size*

$$m \geq \max\left(\frac{4}{\epsilon}\log\left(\frac{2}{\delta}\right), \frac{8\text{VCD}(C)}{\epsilon}\log\left(\frac{13}{\epsilon}\right)\right) \tag{8}$$

and returns any concept consistent with all m examples meets the learning criterion of the pac learning definition.

Technical note on "well-behaved" classes: For anything to work with geometric instances spaces, we must have certain conditions which amount to saying that the concepts are not measure-theoretically pathological. For instance, we certainly require that every concept is a Borel set. A fair amount

[§] Unless explicitly stated otherwise, all logarithms in this work are base 2.

has been written about this issue [5, 6, 8]. In practice, this is a nonissue. As computer scientists, we are really interested only in *exceedingly* well-behaved concept classes. After all, our end goal is to have all these problems analyzed by real-world finite-precision computers. (There may be algorithms where we have to be concerned with round-off errors and numerical stability, but that is a completely separate issue.)

Corollary 1 *Let* $\mathbf{C} = \{(X_n, \mathcal{C}_n)\}_{n\geq 1}$ *be a concept class such that* $\text{VCD}(\mathcal{C}_n) = p(n)$ *for some polynomial* p. *Then* $\mathbf{C}$ *is statistically pac learnable, with a sample size of*

$$m \geq \max\left(\frac{4}{\epsilon}\log\left(\frac{2}{\delta}\right), \frac{8p(n)}{\epsilon}\log\left(\frac{13}{\epsilon}\right)\right). \tag{9}$$

To go from statistical pac learning to true pac learning, it is sufficient to be able to find a concept consistent with a sample in polynomial time. For instance, for the class of axis-parallel boxes mentioned above, this is a simple task. Simply take a sample of the size specified by Corollary 1, and return the smallest axis-parallel boxes containing all the positive instances seen, which can easily be done in time linear in the number of points.

Much of the labor of pac learning researchers consists of searching for clever ways to find a concept consistent with a sample. Unfortunately, for many interesting classes, finding a concept consistent with a sample is NP-complete (which most computer scientists believe means that it is impossible to find a polynomial-time algorithm for this problem). Theorem 2 can be strengthened somewhat to show that with a slightly larger sample size it is sufficient to find a concept that is almost consistent with a sample. In theory, this sometimes makes the problem of finding an acceptable concept much easier computationally, although in practice developing algorithms to find almost consistent concepts is typically still very difficult.

9.3 Lower bounds on sample complexity

The VC dimension also gives us a sort of lower bound on sample complexity for pac learning [10].

Theorem 3 *Fix a concept class* $\mathbf{C} = \{(X_n, \mathcal{C}_n)\}_{n\geq 1}$, *such that* $|\mathcal{C}_n| \geq 3$ *for sufficiently large* n. *Consider any (possibly randomized) algorithm that works* *by calling* EXAMPLES $m = f(\epsilon, \delta, n)$ *times for some* f, *and returning any concept consistent with those* m *examples. To meet the pac learning criterion, it must be that*

$$m = \Omega\left(\frac{1}{\epsilon}\ln\left(\frac{1}{\delta}\right) + \frac{\mathrm{VCD}(\mathcal{C}_n)}{\epsilon}\right).$$

Theorem 3 would seem to imply that we must always have a finite VC dimension in order to pac learn. However, this is not strictly true. The limitations imposed by Theorem 3 can be overcome by algorithms with slightly different behavior. For instance, the class of all finite unions of closed intervals in $[0, 1]$ has infinite VC dimension, but does have a pac learning algorithm [19].[5] The algorithm behaves roughly as follows. It first draws a small sample from EXAMPLES, and makes as its first hypothesis some "small" concept (one that is a union of as few intervals as possible) consistent with that sample. Then another sample is drawn from EXAMPLES, and the initial hypothesis is checked for consistency with it. If the initial hypothesis is not consistent that sample, then the algorithm starts over again, but drawing more examples. This process is repeated as many times as necessary until a hypothesis is found that is consistent with a subsequently drawn sample.[6]

10 Noisy data

So far, we have assumed that all our training data provided by the box EXAMPLES is perfect. In the real world, of course, there is always some noise.

Noise is modeled by replacing the box EXAMPLES with some other box which provides noisy examples. In theoretical computer science jargon, such boxes are usually called "oracles." One can think of these other oracles as first getting an example from EXAMPLES, and then adding noise to it. The output of the noise oracles is all that the learner can observe.

[5]This is an example of using concept complexity, rather than instance space dimension, as the measure of problem size, as was discussed in Section 2.2.

[6]Technically, this algorithm does not meet our definition of pac learning, since it obviously does not always halt in polynomial time. It does, however, have halt in expected polynomial time. See Haussler et al. [13] for discussion about what difference this small perturbation in the definition of pac learning might make.

Three oracles have been studied so far that are relevant to geometric instance spaces (as have several others that apply only to the instance space $\{0,1\}^n$ [12, 24]). The "desired," noiseless output of each oracle is a correctly labeled example (x, s), where x is drawn according to $\mathbf{P}$. We now now give a precise description of the actual outputs from the following noise oracles: MAL_ν [27], RMC_ν [2], MMC_ν [24, 25].

- When MAL_ν is called, with probability $1-\nu$, it does indeed return a correctly labeled (x, s) where x is drawn according to $\mathbf{P}$. With probability ν it returns an example (x, s) about which no assumptions whatsoever may be made. In particular, this example may be maliciously selected by an adversary who has infinite computing power, and has knowledge of the target concept, $\mathbf{P}$, ν, and the internal state of the algorithm calling this oracle. This *malicious noise* oracle models the situation where the learner usually gets a correct example, but some small fraction ν of the time the learner gets noisy examples, and the nature of the noise is unknown or unpredictable.

- When RMC_ν is called, it calls EX to obtain some (noiseless) (x, s), and with probability $1-\nu$, RMC_ν returns (x, s). However, with probability ν, RMC_ν returns $(x, \overline{s})$. This *random misclassification noise* oracle models a benign form of misclassification noise.

- When MMC_ν is called, it also calls EX to obtain some (noiseless) (x, s), and with probability $1-\nu$, MMC_ν returns (x, s). With probability ν, MMC_ν returns (x, l) where l is a label about which no assumption whatsoever may be made. As with MAL_ν we assume an omnipotent, omniscient adversary; but in the case the adversary gets to choose only the label of the example. This *malicious misclassification noise* oracle models a situation in which the only source of noise is misclassification, but the nature of the misclassification is unknown or unpredictable.

10.1 Malicious noise

Not surprisingly, giving an all powerful adversary the ability to tamper with our training data restricts our ability to learn. Kearns and Li [17] showed that, given certain very minimal assumptions, it is impossible to pac learn with examples from MAL_ν unless $\nu < (\epsilon/1 + \epsilon)$. The proof is information-theoretic, meaning that it is impossible even to statistically pac learn under such circumstances.

A generalization of Theorem 2 shows that for very low noise rates, it is sufficient to draw a polynomial-size sample from MAL_ν and return any concept in the concept class that has a minimum disagreement rate with the sample. (The disagreement of a concept with respect to a sample is the number of instances classified differently by the sample and the oracle which provided the examples.)

There is, however, a better method. Any pac learning algorithm for a concept class **C** can be converted to be robust against a malicious noise with a noise rate up to

$$O\left(\frac{\epsilon}{\mathrm{VCD}(\mathbf{C})}\log\left(\frac{\mathrm{VCD}(\mathbf{C})}{\epsilon}\right)\right).$$

The key idea of this meta-algorithm [17] is as follows. The meta-algorithm obtains many, many examples from MAL_ν. These are broken up into blocks of length s, where s is the number of noise-free examples the regular pac learning algorithm requires for **C**. Each block is fed to the regular algorithm, and for each it makes a hypothesis. The number of blocks is chosen so that with high probability one of the blocks is noise-free. One final set of examples is drawn, and the meta-algorithm tests each hypothesis of the regular algorithm on this set, choosing the hypothesis with the lowest disagreement rate.

10.2 Labeling noise

Angluin and Laird studied the case where the only source of noise is random mislabeling [2]. They showed that in this case, for any noise rate strictly less than one half, it is sufficient to pick out the concept with minimum disagreement rate, or even any concept with low disagreement rate. This solves the problem of statistical pac learning. These results extend to the case where the label is maliciously chosen by a powerful adversary (i.e., where RMC_ν is replaced by MMC_ν) [24, 25].

Unfortunately, in this case no meta-algorithm for converting any pac learning algorithm into an algorithm robust against large amounts of mislabeling noise is known. Angluin and Laird do give a particularly clever algorithm for finding a low disagreement rate concept for the concept class

k-CNF (a particular class of boolean formulas on $\{0,1\}^n$).

11 Research directions

Perhaps the most important research goal right now should be to find more pac learning algorithms for more concept classes, concentrating on true pac learning, not statistical pac learning. The most notorious open class is arbitrary DNF formulas on $\{0,1\}^n$. Either an algorithm, or a prove that finding one is impossible (or NP-hard, or equivalent to factoring or what have you) would be of great interest. Of course, it would be nicer to discover a good algorithm than to discover a hardness proof.

Another important open class involves unions of boxes. Algorithms exist for learning arbitrary unions of axis-parallel rectangles in E^2, and arbitrary nested differences of axis-parallel boxes in E^n [8, 15]. Arbitrary unions of boxes (regardless of whether they are axis-parallel) in E^n cannot be pac learnable since the VC dimension is not polynomial in n. It is open whether unions of up to $O(n)$ axis-parallel boxes (or even of $O(\log n)$) in E^n can be pac learned.[7] This concept class is of particular interest because it is also sometimes used in artificial intelligence approaches to data classification (e.g., [23]).

I believe that in the end, the number of good polynomial-time pac learning algorithms found over the intermediate-term future will determine whether this whole line of research continues to flourish, or becomes just one more interesting little side area studied for some period of time by theoretical computer science.

Nevertheless, there is room for a lot of work in the area of noise and geometry. The malicious noise oracle certainly is sufficient to model any conceivable real-world noise, but it is so strong that, as we saw, only limited results are possible. Other than that model, none of the noise models studied to date seems particularly compelling for geometric domains. It would be interesting to study what happens if, say, labels are not affected at all by noise, and instances are perturbed by noise with a normal distribution.

[7] Note that this is related to the problem of DNF boolean formulas mentioned above. A union of n boxes in E^n would generalize learning n-term DNF in $\{0,1\}^n$. This problem is known to be NP-complete if we demand that the output be a union of a *minimum* number of boxes [21]

References

[1] Dana Angluin. Queries and concept learning. *Machine Learning*, 2:319–342, 1987.

[2] Dana Angluin and Philip Laird. Learning from noisy examples. *Machine Learning*, 2(4):343–370, 1988.

[3] Dana Angluin and Carl H. Smith. Inductive inference: Theory and methods. *Computing Surveys*, 15(3):237–269, September 1983.

[4] Patrick Assouad. Densité et dimension. *Ann. Inst. Fourier, Grenoble*, 33(3):233–282, 1983.

[5] S. Ben-David, A. Itai, and E. Kushilevitz. Learning by distances. In *Proc. 3rd Annu. Workshop on Comput. Learning Theory*, pages 232–245, San Mateo, CA, 1990. Morgan Kaufmann.

[6] Shai Ben-David and Gyora M. Benedek. Measurability constraints on pac learnability. Technical report, Technion, Haifa, 1991.

[7] Anselm Blumer, Andrzej Ehrenfeucht, David Haussler, and Manfred K. Warmuth. Occam's razor. *Information Processing Letters*, 24:377–380, April 1987.

[8] Anselm Blumer, Andrzej Ehrenfeucht, David Haussler, and Manfred K. Warmuth. Learnability and the Vapnik-Chervonenkis dimension. *Journal of the ACM*, 36(4):929–965, 1989.

[9] R. M. Dudley. A course on empirical processes. In *Lecture Notes in Mathematics No. 1097*. Springer-Verlag, 1984.

[10] Andrzej Ehrenfeucht, David Haussler, Michael Kearns, and Leslie Valiant. A general lower bound on the number of examples needed for learning. *Information and Computation*, 82:247–261, 1989.

[11] E. Mark Gold. Language identification in the limit. *Information and Control*, 10:447–474, 1967.

[12] Sally A. Goldman and Robert H. Sloan. Can PAC learning algorithms tolerate random attribute noise? *Algorithmica*, 1995. To appear. Preliminary version available as "The Difficulty of Random Attribute Noise," Washington University Dept. of Computer Science Tech. Report WUCS-91-92, 1991.

[13] David Haussler, Michael Kearns, Nick Littlestone, and Manfred K. Warmuth. Equivalence of models for polynomial learnability. *Information and Computation*, 95(2):129–161, December 1991.

[14] David Haussler and Elmo Welzl. Epsilon-nets and simplex range queries. *Discrete Computational Geometry*, 2:127–151, 1987.

[15] David Helmbold, Robert Sloan, and Manfred K. Warmuth. Learning nested differences of intersection-closed concept classes. *Machine Learning*, 5(2):165–196, 1990.

[16] David Helmbold, Robert Sloan, and Manfred K. Warmuth. Learning integer lattices. *SIAM Journal on Computing*, 21(2):240–266, 1992.

[17] M. Kearns and M. Li. Learning in the presence of malicious errors. *SIAM J. Comput.*, 22:807–837, 1993.

[18] Michael Kearns, Ming Li, Leonard Pitt, and Leslie Valiant. Recent results on boolean concept learning. In *Proceedings of the Fourth International Workshop on Machine Learning*, pages 337–352, University of California, Irvine, June 1987.

[19] Nathan Linial, Yishay Mansour, and Ronald L. Rivest. Results on learnability and the Vapnik-Chervonenkis dimension. *Information and Computation*, 90(1):33–49, January 1991.

[20] Judea Pearl. On the connection between the complexity and credibility of inferred models. *Journal of General Systems*, 4:255–264, 1978.

[21] Leonard Pitt and Leslie G. Valiant. Computational limitations on learning from examples. *Journal of the ACM*, 35(4):965–984, 1988.

[22] David Pollard. *Convergence of Stochastic Processes*. Springer-Verlag, 1984.

[23] Steven Salzberg. A nearest hyperrectangle learning method. *Machine Learning*, 6(3):251–276, 1991.

[24] Robert H. Sloan. Four types of noise in data for PAC learning. *Information Processing Letters*. To appear.

[25] Robert H. Sloan. Types of noise in data for concept learning. In *First Workshop on Computational Learning Theory*, pages 91–96. Morgan Kaufmann, 1988.

[26] Leslie G. Valiant. A theory of the learnable. *Communications of the ACM*, 27(11):1134–1142, November 1984.

[27] Leslie G. Valiant. Learning disjunctions of conjunctions. In *Proceedings IJCAI-85*, pages 560–566. International Joint Committee for Artificial Intelligence, Morgan Kaufmann, August 1985.

[28] V. N. Vapnik. *Estimation of Dependences Based on Empirical Data.* Springer-Verlag, New York, 1982.

[29] V. N. Vapnik and A. Ya. Chervonenkis. On the uniform convergence of relative frequencies of events to their probabilities. *Theory of Probability and its Applications*, XVI(2):264–280, 1971.

A Review of Some Extensions to the PAC Learning Model*

Sanjeev R. Kulkarni

Department of Electrical Engineering
Princeton University
Princeton, NJ 08544
kulkarni@ee.princeton.edu

Abstract

The Probably Approximately Correct (PAC) learning model, which has received much attention recently in the machine learning community, attempts to formalize the notion of learning from examples. In this paper, we review several extensions to the basic PAC model with a focus on the information complexity of learning. The extensions discussed are learning over a class of distributions, learning with queries, learning functions, and learning from generalized samples.

*This work was supported in part by the U.S. Army Research Office under grants DAAL03-86-K-0171 and DAAL03-92-G-0320 and by the National Science Foundation under grant IRI-9457645.

1 Introduction

Recently, there has been a great deal of work on the Probably Approximately Correct (or PAC) learning model, which is a precise framework for studying non-parametric function estimation from random samples. Many early and fundamental results on PAC-like models appear in the work of Vapnik [57], and many important results relevant to the PAC model have been obtained in the probability and statistics literature by Vapnik and Chervonenkis [55, 56, 57], Dudley [12], Pollard [42], and others. Valiant [54] independently introduced a learning model in the computer science literature with the additional concern of computational complexity of the learning algorithm. This paper spawned a great deal of work in the machine learning community on analyzing and extending Valiant's model, which is commonly referred to as the PAC model. Recently, Haussler [23] formulated a very general framework refining and consolidating much of the previous work on the PAC model. Much of the literature in the computer science community on PAC learning takes variations of Valiant's formulation [54] as the starting point, and many extensions have been carried out only for variations of this formulation. Hence, we will also take as our starting point the model of Valiant, referring to this model as the basic PAC model.

The basic PAC model can be considered as an attempt to capture the notion of "learning from examples". The essential idea consists of approximating an unknown "concept" from a finite number of positive and negative "examples" of the concept. For example, the concept might be some unknown geometric figure in the plane with the positive and negative examples corresponding to points inside and outside the figure, respectively. The goal is to approximate the figure from a finite number of such points. The examples are assumed to be drawn from some probability distribution, and the same distribution is used to evaluate how well a concept is learned. However, no assumptions are made about which particular distribution is used. That is, the learning is required to take place for every distribution. Since the examples are drawn randomly, the learner is actually only required to approximate the concept with high probability (hence the name PAC).

Descriptions of the basic PAC model have appeared in many places, and a result characterizing learnability for this model in terms of the VC dimension of the concept class is widely known [11]. This result ensures that if the concept class has finite VC dimension then we will have the strong performance guarantees of PAC learnability. However, in some non-parametric settings, there is no learning rule that satisfies the PAC criteria, but one can find learning rules that satisfy other weaker but still desirable criteria. There are also many problems in which the learner receives information other than just random examples. For example the learner may have some control over the examples received, or in fact may receive data other than examples. Furthermore, in many problems the "concept" to be learned may more naturally be thought of as a function defined on some domain rather than a subset (binary-valued function) of a domain. Thus, it is natural to consider extensions of this basic PAC model which relax the learning requirements, provide different information gathering mechanisms, and generalize the notion of concepts to be learned.

In this paper, we review some results on several extensions of the basic PAC model. Our focus is on the information complexity of learning, and accordingly we ignore issues of computational complexity. Also, although a statistical framework for the observations is a more general and natural one (e.g., see [23, 34]), following much of the work in the PAC learning community we focus here on the "noise-free" case. For completeness in Section 2 we give a precise definition of the framework and state the well known result characterizing learnability in terms of the VC dimension of the concept class. In Section 3, we discuss the problem of learning over a class of distributions which has been considered in [7, 34, 40]. Some results from [16, 31] on learning with queries are discussed in Section 4. The problem of learning real-valued functions (a special case of a formulation in [23]) is discussed in Section 5, and an extension of this framework to learning from generalized samples [32] is discussed in Section 6. Throughout these sections we mention a number of open problems. Finally, in Section 7 we briefly mention some other directions which have been considered.

2 Basic PAC Learning Model

Informally, the learning framework can be described as follows. The *learner* wishes to learn a concept unknown to him. The *teacher* provides the learner with random positive and negative examples of the concept drawn according to some probability distribution. From a finite set of examples, the learner outputs a hypothesis which is his current estimate of the concept. The error of the estimate is taken as the probability that the hypothesis will incorrectly classify the next randomly chosen example. The learner cannot be expected to exactly identify the concept since only a finite number of examples are seen. Also, since the examples are randomly chosen, there is some chance that the hypothesis will be very far off (due to poor examples). Hence, the learner is only required to closely approximate the concept with sufficiently high probability from some finite number of examples. Furthermore, the number of examples required for a given accuracy and confidence should be bounded independent of the distribution from which the examples are drawn. Below, we will describe this framework precisely, following closely the notation of [11].

Let X be a set which is assumed to be fixed and known. X is sometimes called the *instance space*. Typically, X is taken to be either $\mathbf{R}^n$ (especially $\mathbf{R}^2$) or the set of binary n-vectors. A *concept* will refer to a subset of X, and a collection of concepts $C \subseteq 2^X$ will be called a *concept class.* An element $x \in X$ will be called a *sample*, and a pair $\langle x, a\rangle$ with $x \in X$ and $a \in \{0,1\}$ will be called a *labeled sample.* Likewise, $\overline{x} = (x_1, \ldots, x_m) \in X^m$ is called an *m-sample*, and a *labeled m-sample* is an m-tuple $(\langle x_1, a_1\rangle, \ldots, \langle x_m, a_m\rangle)$ where $a_i = a_j$ if $x_i = x_j$. For $\overline{x} = (x_1, \ldots, x_m) \in X^m$ and $c \in C$, the *labeled m-sample of c generated by* $\overline{x}$ is given by $sam_c(\overline{x}) = (\langle x_1, I_c(x_1)\rangle, \ldots, \langle x_m, I_c(x_m)\rangle)$ where $I_c(\cdot)$ is the indicator function for the set c. The *sample space* of C is denoted by S_C and consists of all labeled m-samples for all $c \in C$, all $\overline{x} \in X^m$, and all $m \geq 1$.

Let H be a collection of subsets of X. H is called the *hypothesis class*, and the elements of H are called *hypotheses.* Let $\mathcal{A}_{CH}$ be the set of all functions $A : S_C \to H$. A function $A \in \mathcal{A}_{CH}$ is called *consistent* if it always produces a hypothesis which agrees with the samples, i.e. whenever $h = A(\langle x_1, a_1\rangle, \ldots, \langle x_m, a_m\rangle)$ we have $I_h(x_i) = a_i$ for $i = 1, \ldots, m$. Given a probability distribution P on X, the *error* of A with respect to P for a concept $c \in C$ and sample $\overline{x}$ is defined as $error_{A,c,P}(\overline{x}) = P(c\Delta h)$ where $h = A(sam_c(\overline{x}))$ and $c\Delta h$ denotes the symmetric difference of the sets c and h. Finally, in the definition of learnability to be given below, the samples used in forming a hypothesis will be drawn from X independently according to the same probability measure P. Hence, an m-sample will be drawn from X^m according to the product measure P^m.

We can now state the following definition of distribution-free learnability (or learnability for every distribution), which is the version from Blumer et al. [11] of Valiant's [54] definition (without restrictions on computability – see below).

Definition 1 (Distribution-free Learning) *The pair (C, H) is* learnable *if there exists a function $A \in \mathcal{A}_{CH}$ such that for every $\epsilon, \delta > 0$ there is a $0 < m < \infty$ such that for every probability* *measure P and every $c \in C$, if $\overline{x} \in X^m$ is chosen at random according to P^m then the probability that $error_{A,c,P}(\overline{x}) < \epsilon$ is greater than $1 - \delta$.*

Several comments concerning this definition are in order. First, learnability depends on both the concept class C and the hypothesis class H, which is why we defined learnability in terms of the pair (C, H). However, in the literature the case $H = C$ is often considered, in which case, for convenience, we may speak of learnability of C in place of (C, C). In fact, we will be considering the case that $H = C$ throughout. Second, the sample size m is clearly a function of ϵ and δ but a fixed $m = m(\epsilon, \delta)$ must work uniformly for every distribution P and concept $c \in C$, which is the reason for the term *distribution-free* learning. Finally, ϵ can be thought of as an accuracy parameter while δ can be thought of as a confidence parameter. The definition requires that the learning algorithm f output a hypothesis that with high probability (greater than $1 - \delta$) is approximately correct (to within ϵ), which gave rise to the term the term *probably approximately correct* (PAC) learning to describe this definition [3].

A somewhat more general and useful definition of learnability was actually used by in [54] and later by others. This definition incorporates both a notion of the size or complexity of concepts and the central idea that the learning algorithm (i.e., the function which produces a hypothesis from labeled samples) should have polynomial complexity in the various parameters. Other variations of this definition, such as seeing positive examples only, or having the choice of positive or negative examples, have also been considered. Some equivalences among the various definitions learnability were shown in [21]. In this paper, we will not consider these variations, and as mentioned before we will be concerned only with the information complexity of learning and so will not consider issues of the computational complexity of learning algorithms.

A fundamental result relates learnability for every distribution to a combinatorial parameter of the concept class called the Vapnik-Chervonenkis (VC) dimension The notion of VC dimension was introduced in [55] and has since been studied and used by a number of authors (e.g., see [12]).

Definition 2 (Vapnik-Chervonenkis Dimension) *Let $C \subseteq 2^X$. For any finite set $S \subseteq X$, let $\Pi_C(S) = \{S \cap c : c \in C\}$. S is said to be* shattered *by C if $\Pi_C(S) = 2^S$. The* Vapnik-Chervonenkis dimension *of C is defined to be the largest integer d for which there exists a set $S \subseteq X$ of cardinality d such that S is shattered by C. If no such largest integer exists then the VC dimension of C is infinite.*

The following theorem shows that the VC dimension of a concept class characterizes its learnability. This theorem is a special case of results of Vapnik [57], but was also proved by Blumer et al. [11] using results from [55] in the context of PAC learning. The sample size bounds shown reflect some refinements due to Ehrenfeucht et al. [15] and Blumer et al. [11].

Theorem 1 *A concept class C (subject to certain measurability conditions) is learnable for every distribution iff the VC dimension of C is finite. Furthermore, if d is the VC dimension of C then a sample size $O(\frac{1}{\epsilon}\log\frac{1}{\delta} + \frac{d}{\epsilon}\log\frac{1}{\epsilon})$, is sufficient and for $\epsilon < 1/2$ a sample size $\Omega(\frac{1}{\epsilon}\log\frac{1}{\delta} + \frac{d}{\epsilon})$ is necessary.*

In the following sections, we review some extensions of the basic PAC model and discuss results analogous to Theorem 1 characterizing learnability and providing sample size bounds for the various extensions.

3 Learning Over a Class of Distributions

In distribution-free learning, there must exist a sample size which works uniformly for all distributions. One motivation for this requirement is that the learner often does not know and cannot control the distribution used to draw the samples. The uniform sample size requirement guarantees that the learner can be successful even in this case. A natural question is to study how much can be gained if the learner has prior knowledge as to the distribution used. Specifically, suppose the learner knows that the distribution P comes from a class of distributions $\mathcal{P}$, and that learning is required to take place uniformly only for $P \in \mathcal{P}$. The definition of learnability in this case is completely analogous to the definition given earlier, but for completeness we state it formally.

Definition 3 (Learnability for a Class of Distributions) *Let $\mathcal{P}$ be a fixed and known collection of probability measures. The pair (C, H) is said to be* learnable with respect to $\mathcal{P}$ *if there exists a mapping $A \in \mathcal{A}_{CH}$ such that for every $\epsilon, \delta > 0$ there is a $0 < m < \infty$ such that for every probability measure $P \in \mathcal{P}$ and every $c \in C$, if $\overline{x} \in X^m$ is chosen at random according to P^m then the probability that $error_{A,c,P}(\overline{x}) < \epsilon$ is greater than $1 - \delta$.*

As before, we will be consider the case $H = C$ throughout, so that we will simply speak of learnability of C rather than learnability of (C, H).

One extreme case is when $\mathcal{P}$ is the set of all probability distributions over some fixed σ-algebra of X (which we will denote by $\mathcal{P}^*$). Then the above definition reduces to learnability for all distributions (distribution-free learning). As discussed in the previous section, the VC dimension of C characterizes learnability in this case.

The other extreme case is learnability for a fixed distribution, namely when $\mathcal{P}$ consists of a single distribution P (i.e., $\mathcal{P} = \{P\}$). Conditions for learnability in this case were studied by Benedek and Itai [7] in terms of covering numbers/metric entropy of the concept class with respect to the fixed distribution.

Definition 4 Covering Number and Metric Entropy *Let (Y, ρ) be a metric space. Define $N(\epsilon) \equiv N(\epsilon, Y, \rho)$ to be the smallest integer n such that there exists $y_1, \ldots, y_n \in Y$ with $Y = \cup_{i=1}^{n} B_\epsilon(y_i)$ where $B_\epsilon(y_i)$ is the open ball of radius ϵ centered at y_i. If no such n exists, then $N(\epsilon, Y, \rho) = \infty$. $N(\epsilon)$ is called the* covering number *of Y and $\log_2 N(\epsilon)$ is called the* metric entropy *of Y*

$N(\epsilon)$ represents the smallest number of balls of radius ϵ which are required to cover Y. For another interpretation, suppose we wish to approximate Y by a finite set of points so that every element of Y is within ϵ of at least one member of the finite set. Then $N(\epsilon)$ is the smallest number of points possible in such a finite approximation of Y. The notions of covering numbers and metric entropy for various metric spaces have been studied and used by a number of authors (e.g. see [12, 13, 28, 53]). These notions are equally valid in the case where ρ is only a pseudo-metric rather than a metric. (A pseudo-metric satisfies the requirements of a metric except that $\rho(y_1, y_2) = 0$ need not imply $y_1 = y_2$, i.e., it may happen that $\rho(y_1, y_2) = 0$ even though $y_1 \neq y_2$.)

For the learning problem under consideration, the measure of error $d_P(c_1, c_2) = P(c_1 \Delta c_2)$ between two concepts c_1, c_2 with respect to a distribution P is a pseudo-metric. Note that $d_P(\cdot, \cdot)$ is only a pseudo-metric since c_1 and c_2 may be unequal but may differ on a set of measure zero with respect to P, so that $d_P(c_1, c_2) = 0$.

For convenience, if P is a distribution we will use the notation $N(\epsilon, C, P)$ (instead of $N(\epsilon, C, d_P)$), and we will speak of the metric entropy of C with respect to P, with the understanding that the (pseudo-) metric being used is $d_P(\cdot, \cdot)$.

Benedek and Itai [7] showed the following result.

Theorem 2 *A concept class C is learnable with respect to a fixed distribution P iff C has finite metric entropy with respect to P for all $\epsilon > 0$. Furthermore, a sample size $(32/\epsilon)\ln(N(\epsilon/2)/\delta)$ is sufficient and a sample size sample size $\log_2[(1-\delta)N(2\epsilon)]$ is necessary for ϵ, δ learnability.*

As noted in [7], a function that simply returns some concept consistent with the samples does not necessarily learn. For example, let $X = [0, 1]$, P be the uniform distribution on X, and C be the concept class containing all finite sets of points and the entire unit interval. That is,

$$C = \{[0,1]\} \cup \{\{x_1, \ldots, x_r\} : r < \infty, x_i \in [0,1] \ \forall i\}$$

If the target concept is $[0, 1]$ then for every finite sample there are many concepts which are consistent with the sample but are ϵ-far (with $\epsilon = 1$) from $[0, 1]$. Namely, any finite set of points which contains the points of the sample is a concept with this property.

The similarity of Theorems 1 and 2 naturally suggest that there may be some relationships between the VC dimension of a concept class and its metric entropy with respect to various distributions. This is indeed the case. The known relationships essentially provide upper and lower bounds to $\sup_P N(\epsilon, C, P)$ in terms of the VC dimension of the concept class. Upper bounds are more difficult to obtain since these require a uniform bound on the metric entropy over all distributions. The lower bounds result from statements of the form that there exists a distribution P (typically a uniform distribution over some finite set of points) for which $N(\epsilon, C, P)$ is greater than some function of the VC dimension. The following theorem summarizes some known upper and lower bounds.

Theorem 3 *Let C be a concept class with infinitely many concepts and let $1 \le d < \infty$ be the VC dimension of C. For $\epsilon \le 1/2$,*

$$\sup_P \log_2 N(\epsilon, C, P) \ \ge \ \max\{2d(1/2 - 2\epsilon)^2 \log_2 e, \log_2 \frac{1}{2\epsilon}\}$$

and for $\epsilon \le 1/2d$,

$$\sup_P \log_2 N(\epsilon, C, P) \le d \log_2 \left(\frac{2e}{\epsilon} \ln \frac{2e}{\epsilon}\right) + 1$$

The first and second terms of the lower bound are from [29, 31]. The upper bound is from [23] which is a refinement of a result from [42] using techniques originally from [12]. A weaker upper bound was also given in [7]. Note that the condition that C contain infinitely many concepts is required only for the second term of the lower bound. Also, note that as an immediate corollary of the theorem we have the result that $\sup_P N(\epsilon, C, P) < \infty$ iff the VC dimension of C is finite.

The similarity of Theorems 1 and 2 together with Theorem 3 suggest that perhaps the condition

$$N(\epsilon, C, \mathcal{P}) \equiv \sup_{P \in \mathcal{P}} N(\epsilon, C, P) < \infty$$

for every $\epsilon > 0$ is necessary and sufficient for learnability of C with respect to $\mathcal{P}$, as conjectured in [7]. Necessity is obvious, but it has recently been shown that this condition is not sufficient in general for learnability [14].

Results for some specific classes of distributions were obtained by Kulkarni [29], and these results have recently been extended by Kulkarni and Vidyasagar [34]. For example, a natural extension to considering a single distribution P_0 is to consider the class of all distributions sufficiently close to P_0, i.e., "spheres" of distributions in an appropriate sense. Two types of spheres of distribution were considered in [29, 34]. Specifically, for a given distribution P_0 and $0 \leq \lambda \leq 1$ define

$$\mathcal{P}_v(P_0, \lambda) = \{P \in \mathcal{P}^* : \|P - P_0\| \leq \lambda\}$$

where

$$\|P_1 - P_2\| = \sup_{\text{measurable } A} |P_1(A) - P_2(A)|$$

is total variation between P_1 and P_2, and as before $\mathcal{P}^*$ is the set of all probability measures over some fixed σ-algebra. $\mathcal{P}_v(P_0, \lambda)$ represents all probability measures which are within λ of P_0 in total variation.

Alternatively, note that if P_1 and P_2 are probability measures then the convex combination $\lambda P_1 + (1-\lambda)P_2$ for $0 \leq \lambda \leq 1$ is also a probability measure. Given a distribution P_0 and $0 \leq \lambda \leq 1$, define

$$\mathcal{P}_l(P_0, \lambda) = \{(1-\eta)P_0 + \eta P : \eta \leq \lambda, P \in \mathcal{P}^*\}$$

The distributions in $\mathcal{P}_l(P_0, \lambda)$ can be thought of as those obtained by using P_0 with probability greater than or equal to $1 - \lambda$ and using an arbitrary distribution otherwise. Note that, we have $\mathcal{P}_v(P_0, 0) = \mathcal{P}_l(P_0, 0) = \{P_0\}$ and $\mathcal{P}_v(P_0, 1) = \mathcal{P}_l(P_0, 1) = \mathcal{P}^*$.

It was shown in [29, 34] that for $\mathcal{P}_v(P_0, \lambda)$ and $\mathcal{P}_l(P_0, \lambda)$, a concept class is learnable over these spheres of distributions iff it is learnable over all distributions, and hence iff it has finite VC dimension. Furthermore, there is not a significant gain in the sample complexity of learning. Thus, this type of prior knowledge regarding the distribution does not help significantly. The result also suggests that attempts at improving learnability by using density estimation to extract information as to the distribution being used will also not be particularly useful.

Another interesting class of distributions considered in [29, 34] was the set of mixture distributions obtained from a finite set of distributions $P_1, \ldots, P_n$. Specifically, given distributions $P_1, \ldots, P_n$, define $conv(P_1, \ldots, P_n)$ as

$$conv(P_1, \ldots, P_n) \quad = \quad \{\lambda_1 P_1 + \cdots + \lambda_n P_n : \ \lambda_i \geq 0, \ \lambda_1 + \cdots + \lambda_n = 1\}$$

This is the set of all distributions that can be written as a convex combination of $P_1, \ldots, P_n$. For such sets it was also shown that a concept class is learnable with respect to $conv(P_1, \ldots, P_n)$ iff it is learnable with respect to each P_i individually. Thus, prior knowledge that the distribution is mixture of a finite number of distributions can help significantly in terms of learnability.

It was also shown in [29, 34] that learnability of a concept class is retained under finite unions of classes of distributions. That is, if a concept class C is learnable for a finite number of sets of distributions $\mathcal{P}_1, \ldots, \mathcal{P}_n$ then it is learnable with respect to their union $\mathcal{P} = \cup_{i=1}^n \mathcal{P}_i$. Thus, results shown for classes of distributions can be extended to finite unions of such classes.

Some results regarding learnability with respect to a general class of distributions are known. First, as previously mentioned, it is easy to show that the condition $\sup_{P \in \mathcal{P}} N(\epsilon, C, P) < \infty$ is necessary for learnability. Natarajan [40] has shown that a somewhat different condition is sufficient. As before, given a finite set of points $x_1, \ldots, x_l$ let $\Pi_C(x_1, \ldots, x_l)$ denote the subsets of $x_1, \ldots, x_l$ generated by intersection with a member of C. Let $|\Pi_C(x_1, \ldots, x_l)|$ denote the cardinality of $\Pi_C(x_1, \ldots, x_l)$, and let $H_{P,l}(C) = \log_2 E_P|\Pi_C(x_1, \ldots, x_l)|$ where E_P denotes the expected value with the points $x_1, \ldots, x_l$ drawn independently according to P. The quantity $H_{P,l}(C)$ was introduced by Vapnik and Chervonenkis [55] and was referred to as a notion of entropy. Based on the results of Vapnik and Chervonenkis [55, 56], Natarajan has shown that the condition $\lim_{l\to\infty} \sup_{P\in\mathcal{P}} H_{P,l}(C)/l = 0$ is sufficient for learnability of C with respect to $\mathcal{P}$. A simple example shows that this condition is not necessary. Let $X = [0,1]$, P be the uniform distribution on X, and let C consist of the interval $[0,1]$ itself together with all finite sets of points. Then C is learnable since every concept is a distance 0 away from either the empty set or the entire interval, so that letting the hypothesis be either $\emptyset$ or $[0,1]$ based on a single sample is a learning algorithm. On the other hand, $H_{P,l}(C) = 1$ for all l.

Using a distinction considered in [5] and [40] we can refine the problem of learning for a class of distributions. A concept class will be called strongly learnable if the learner can output any concept consistent with the examples and still have a learning algorithm. The term learnable by itself will still refer to the case where some hypothesis (that need not be consistent) is guaranteed to be close to the target concept. (Note that, unfortunately, this terminology is not particularly good as the term "weak learnability" has been used by others to denote a different idea [49].) The result of Natarajan shows that $\lim_{l\to\infty} \sup_{P\in\mathcal{P}} H_{P,l}(C)/l = 0$ is sufficient for strong learnability since the condition is equivalent to a uniform convergence property of the empirical measures [55] which in turn implies strong learnability. However, it is not necessary since the concept class consisting of all finite subsets of $[0,1]$ is strongly learnable under the uniform distribution (all concepts are a distance zero apart) but the condition (and hence uniform convergence) is clearly violated. As far as we know, it is an open problem to find a single condition which is both necessary and sufficient for strong learnability. Similarly, the problem of characterizing (not necessarily strong) learnability, is also open.

4 Learning With Queries

In the basic PAC model the examples are provided to the learner according to some probability distribution over which the learner has no control. In this sense, the model assumes a purely passive learner. There has been quite a bit of work done on studying models of learnability in which the learner has more powerful information gathering mechanisms. Specifically the use of various types of queries has been considered for several learning models. For example Angluin [4] studied a variety of oracles and their effect on learning. Some types of queries considered by Angluin include membership queries, equivalence queries, subset, superset, and disjointness queries, and exhaustiveness queries.

Much of the work on studying the effects of queries has been for inductive inference models of learning with issues of computational complexity (as well as information complexity) playing a central role (e.g., see [19, 27]). Amsterdam [2] considered PAC learning when the learner has access to an "experiment" oracle which gives the learner some control over the choice of examples by allowing the learner to focus attention on chosen regions of the instance space. Again, the concern was with both computational and information complexity. Here we briefly discuss certain results concerning the effect of queries on only the information complexity of PAC learning.

Eisenberg and Rivest [16] studied the effect on the sample complexity of PAC learning by allowing membership queries in addition to random examples. They introduced the following notion of a dense-in-itself concept class.

Definition 5 *A concept class C is called* dense-in-itself *if for all $c \in C$, all $\gamma > 0$, and all finite measures μ on X, there is a concept $c' \in C$ with $c' \neq c$ such that $\mu(c \Delta c') < \gamma$.*

They obtained the following result which shows that allowing membership queries does not significantly improve the sample complexity of learning a dense-in-itself concept class.

Theorem 4 *If C is dense-in-itself then a sample size (i.e., total number of random examples and membership queries) $\Omega((1/\epsilon)\ln(1/\delta))$ is necessary for ϵ,δ learnability.*

As mentioned previously, the sample size needed using just random examples is $\Omega((1/\epsilon)\ln(1/\delta) + d/\epsilon)$ for distribution-free learnability where d is the VC dimension of C. For fixed d, the sample complexities with and without membership queries are within a constant factor of one another (as pointed out in [16]).

In [16], Eisenberg and Rivest also considered learning over a class of "smooth" distributions. They used the following definitions. A distribution P on $\mathbf{R}^n$ is called *smooth* if for some s,

$$P(A) \leq s \cdot \text{volume}(A)$$

for all A. A concept class C is said to *have all error rates* with respect to a distribution P if for every $c \in C$ and every $\alpha < \sup_{c' \in C} d_P(c, c')$ there is a $\hat{c} \in C$ with $d_P(c, \hat{c}) = \alpha$.

They considered learning over a class of smooth distributions with fixed and known smoothness s. They showed that learning a concept class C with all error rates with respect to a smooth distribution P without the use of membership queries requires at least $\frac{1}{2\epsilon}\ln(1/\delta)$ random samples. On the other hand, for the concept class over $[0,1]$ consisting of all intervals of the form $[0,a]$, they showed that for a smooth distribution (which makes this class have all error rates) allowing membership queries reduces the sample complexity to $\ln(s/\epsilon)$. Hence, for this particular concept class allowing membership queries does substantially reduce the sample complexity. They left open the question of whether allowing membership queries results in a similar reduction in sample the complexity for learning more general concept classes over smooth distributions.

Some other work on the effect of queries on the sample complexity of learning was done by Kulkarni et al. [31]. They studied the limits of what can be gained by allowing the most general set of binary valued learner-environment interactions, and giving the learner complete control over the information gathering. Specifically, they considered completely "active" learning in that the the learner is allowed to ask arbitrary yes/no (i.e., binary valued) questions, and these questions need not be decided on beforehand. That is, the questions the learner asks can depend on previous answers and can also be generated randomly. Many of the oracles previously considered in the literature are simply particular types of yes/no questions (although those oracles that provide counterexamples are not).

In active learning, the probability distribution is not used to generate the examples, but is used only to measure the distance between concepts. For a fixed distribution, the only object unknown to the learner is the chosen concept. In this case, an arbitrary binary question provides information of the type $c \in C_0$ where C_0 is some subset of C. That is, all binary questions can be reduced to partitioning C into two pieces and asking to which of the two pieces does c belong. For distribution-free learning (or more generally, learning for a class of distributions) the distribution P is also unknown. In this case, every binary question can be reduced to the form "Is $(c, P) \in q$?" where q is an arbitrary subset of $C \times \mathcal{P}$, so that C and $\mathcal{P}$ can be simultaneously and dependently partitioned. This follows by letting q be the set of (c, P) pairs for which the answer to the binary question is "yes." Thus, the information the active learner obtains is of the form $(\langle q_1, a_1\rangle, \ldots, \langle q_m, a_m\rangle)$ where $q_i \subseteq C \times \mathcal{P}$ and $a_i = 1$ if $(c, P) \in q_i$ and $a_i = 0$ otherwise. The q_i correspond to the binary valued (i.e., yes/no) questions and a_i denotes the answer to the question q_i when the true concept and probability measure are c and P respectively.

In general, q_i can be generated randomly or deterministically and can depend on all previous questions and answers $\langle q_1, a_1\rangle, \ldots, \langle q_{i-1}, a_{i-1}\rangle$. The q_i are not allowed to depend explicitly on the true concept c and probability measure P, but can depend on them implicitly through answers to previous questions. Let $\overline{q} = (q_1, \ldots, q_m)$ denote a set of m questions generated in such a manner, and let $sam_{c,P}(\overline{q})$ denote the set of m question and answer pairs when the true concept and probability measure are c and P respectively. Let $S_{C,\mathcal{P}}$ denote all sets of m question and answer pairs generated in such a manner for all $c \in C$, $P \in \mathcal{P}$, and $m \geq 1$. An active learning algorithm is an algorithm Q for selecting $q_1, \ldots, q_m$ together with a mapping $A : S_{C,\mathcal{P}} \to C$ for generating a hypothesis from $sam_{c,P}(\overline{q})$. In general, Q and/or A may be random, which results in probabilistic active learning algorithms.

Definition 6 (Active Learnability for a Class of Distributions) *Let $\mathcal{P}$ be a fixed and known collection of probability measures. C is said to be* actively learnable with respect to $\mathcal{P}$ *if there exists an algorithm Q for selecting queries and a mapping $A : S_{C,\mathcal{P}} \to C$ such that for every $\epsilon, \delta > 0$ there is a $0 < m(\epsilon, \delta) < \infty$ such that for every probability measure $P \in \mathcal{P}$ and every $c \in C$, if $h = A(sam(c, P))$ then the probability (with respect to any randomness in Q and A) that $P(h \Delta c) < \epsilon$ is greater than $1 - \delta$.*

The following result was obtained in [31] which shows that the as in learning with random examples, finite metric entropy is required in the case of active learning with respect to a fixed distribution as well. Note that active learning for a fixed distribution is essentially an elementary problem in source coding from information theory [18], but in which the learner wants to minimize the length of the longest codeword rather than the mean codeword length.

Theorem 5 *A concept class C is actively learnable with respect to a distribution P iff $N(\epsilon, C, P) < \infty$ for all $\epsilon > 0$. Moreover, $\lceil \log_2((1-\delta)N(2\epsilon, C, P))\rceil$ queries are necessary, and $\lceil \log_2((1-\delta)N(\epsilon, C, P))\rceil$ queries are sufficient. For deterministic learning algorithms, $\lceil \log_2 N(\epsilon, C, P)\rceil$ queries are both necessary and sufficient.*

Thus, compared with passive learning for a fixed distribution, active learning does not enlarge the set of learnable concept classes, but as expected, fewer queries are required in general. Note that very little is gained by allowing the learner to make mistakes with probability δ. That is, there is a very weak dependence on δ in the sample size bounds.

The case of distribution-free active learnability is a little more subtle than active learnability for a fixed distribution. For both active and passive learning, the requirement that the learning be distribution-free imposes two difficulties. The first is that there must exist a uniform bound on the number of examples or queries over all distributions — i.e., a bound independent of the underlying distribution. The second is that the distribution is unknown to the learner, so that the learner does not know how to evaluate distances between concepts. Hence, since the metric is unknown, the learner cannot simply replace the concept class with a finite ϵ-approximation as in the case of a fixed and known distribution.

For passive learnability, the requirement that the concept class have finite VC dimension is necessary and sufficient to overcome both of these difficulties. However, for active learning the second difficulty is severe enough that no learning can take place as long as the concept class is infinite.

Theorem 6 *C is actively learnable for all distributions iff C is finite.*

Essentially, if the distribution is unknown, then the active learner has no idea about "where" to seek information about the concept. On the other hand, in passive learnability the examples are provided according to the underlying distribution, so that information is obtained in regions of importance. Hence, in the distribution-free case, random samples (from the distribution used to evaluate performance) are indispensable.

If the second difficulty is removed by assuming that the learner has knowledge of the underlying distribution, then the distribution-free requirement plays a part only in forcing a uniform bound on the number of queries needed. For this case, the following result was shown in [31].

Theorem 7 *If the learner knows the underlying probability distribution then C is actively learnable for all distributions iff C has finite VC dimension. Moreover, $\lceil \sup_P \log_2((1-\delta)N(2\epsilon, C, P)) \rceil$ questions are necessary and $\lceil \sup_P \log_2((1-\delta)N(\epsilon, C, P)) \rceil$ questions are sufficient. For deterministic algorithms $\lceil \sup_P \log N(\epsilon, C, P) \rceil$ questions are both necessary and sufficient.*

Hence, even with the side information as to the distribution being used, active learning does not enlarge the set of learnable concept classes. However, as before one would expect an improvement in the number of samples required. A comparison can be made using bounds relating the VC dimension of a concept class to its metric entropy with respect to various distributions as discussed in Section 3. Specifically, Theorem 3 provides upper and lower bounds to $\sup_P N(\epsilon, C, P)$. This theorem gives bounds on the number of questions needed in distribution-free active learning (with the side information) directly in terms of ϵ, δ and the VC dimension of C. The bounds as stated in Theorem 3 are directly applicable to deterministic active learning algorithms or for active learning with $\delta = 0$. For probabilistic algorithms with $\delta > 0$ the quantity $\log_2 1/(1-\delta)$ needs to be subtracted from both the lower and upper bounds.

For completeness, we mention that results can also be obtained if the learner is provided with "noisy" answers to the queries. The effects of various types of noise in the basic PAC model have been studied [3, 25, 51]. For active learning, two natural noise models are random noise in which the answer to a query is incorrect with some probability $\eta < 1/2$ independent of other queries, and malicious noise in which an adversary gets to choose a certain number of queries to receive incorrect answers. For random noise, the problem is essentially equivalent to communication through a binary symmetric channel, so that standard results from information theory on the capacity and coding for such channels [18] can be applied. For malicious noise, some results on binary searching with these types of errors [46] can be applied. For both noise models, the conditions for fixed distribution and distribution-free learnability are the same as the noise-free case, but with a larger sample complexity. However, the more interesting aspects of the active learning results are the indications of the limitations of active learning, and the noise-free case makes stronger negative statements.

The results discussed in this section show that, as expected, the use of oracles can often aid in the learning process. However, considering information complexity only, the use of oracles does not enlarge the set of learnable concept classes. Of course, in addition to reducing the information complexity of learning, oracles can also significantly affect the computational complexity of learning. Taking computational considerations into account, the set of learnable concept classes may certainly be enlarged by allowing oracles.

There are a number of open problems related to the results of this section. For example, the case of noisy queries for active learning was briefly mentioned above. One could also investigate the effects of noise on the model studied by Eisenberg and Rivest that we discussed. Also, it may be interesting to study the effect of other oracles (such as some of those considered in [4]) on the sample complexity of learning. One specific open problem that may be interesting to pursue is to study the reduction in sample complexity of distribution-free learning if the learner has access to both random examples and arbitrary yes/no questions. This is similar to the problem considered in [16], but there the learner could choose only examples to be labeled rather than ask arbitrary questions. Our result for the case where the learner knows the distribution being used provides a lower bound, but if the distribution is not known then we expect that for certain concept classes much stronger lower bounds would hold. In particular, we conjecture that results analogous to those in [16] hold in the case of arbitrary binary valued questions, so that, for example, asking yes/no questions could reduce the sample complexity to learn a dense-in-itself concept class (as defined in [16]) by only a constant factor. On the other hand, if learning was required only over smooth distributions, then as suggested in [16] one would expect that membership queries would significantly reduce the sample complexity of learning.

5 Learning Functions

In the learning model discussed in the previous sections, a concept c was a subset of the instance space X. By identifying a subset with its indicator function, the concepts can be thought of as binary valued functions on X. A natural extension is to consider learning more general functions (i.e. not necessarily binary valued) defined on X.

Such a model is actually contained in the scope of Vapnik's work [57]. Some progress towards learning functions also arose independently out of extensions to Valiant's model (e.g., see [41]). A very general framework was formulated by Haussler [23] building on fundamental work by Vapnik and Chervonenkis [55, 56, 57], Dudley [12], and Pollard [42], and others, as well as consolidating some PAC learning extensions of Valiant's work.

In Haussler's formulation, a learning problem is defined by six components $X, Y, A, F, \mathcal{P}$, and L. The components X, Y, and A are arbitrary spaces. The concept class (hypotheses), denoted by F, is a collection of functions from X to A. The samples are drawn according to a distribution $P \in \mathcal{P}$ on $X \times Y$ where $\mathcal{P}$ is some class of distributions. L is a loss function defined on $Y \times A$ and taking values in $\mathbf{R}^+$ (the positive real numbers). The goal of the learner is to produce a hypothesis $h \in F$ which is close to the optimal in the sense of minimizing the expected loss between the prediction of the hypothesis $h(x)$ and y on a random sample (x, y).

Rather than dealing with the general formulation of [23], we will discuss a special case which will greatly simplify the notation and presentation. Specifically, we will assume that $A = Y = \mathbf{R}$ and that the loss function L is just the absolute value. Furthermore, we will assume that, as in the basic PAC model discussed above, there is some target function $f \in F$ and that the distribution P is on X. Hence, F is a collection of real-valued functions on X. A particular $f \in F$ is chosen, and we observe a set of samples where each sample is of the form $(x, f(x))$ with x drawn according to some $P \in \mathcal{P}$. We are interested producing a hypothesis $h \in F$ for which $d_P(f,h) \equiv E_P|f(x) - h(x)|$ is small where E_P denotes the expectation with respect to P. As before, S_F will denote the set of all m-samples over all $f \in F$, and all $m \geq 1$. The definition of learnability is formally stated below.

Definition 7 (Learning Real-Valued Functions) *Let $X, F, \mathcal{P}$ be as defined above. F is said to be* learnable *if there is a mapping $A : S_F \to F$ for producing a hypothesis h from a set of labeled samples such that for every $\epsilon, \delta > 0$ there is a $0 < m = m(\epsilon, \delta) < \infty$ such that for every probability measure $P \in \mathcal{P}$ and every $f \in F$, if h is the hypothesis produced from a labeled m-sample drawn according to P^m then the probability that $d_P(f,h) < \epsilon$ is greater than $1 - \delta$.*

As mentioned above, this definition is a very restricted case of the formulation in [23]. To get learnability results, generally some assumptions need to be made about the functions in F (or in the more general formulation the loss function) which restricts the occurrence of large values. A simple condition is to assume that the functions themselves are bounded, so that for some $M < \infty$, $f : X \to [0, M]$ for every $f \in F$. Furthermore, certain measurability conditions are also required for the functions (or $L(f(x), y)$ in the more general case), but we will not concern ourselves these conditions here and will assume that F satisfies the required conditions (e.g., see [23]).

For a fixed distribution, the notion of metric entropy provides a sufficient condition for learning as stated in the following theorem.

Theorem 8 *A collection of functions F with $0 \leq f \leq M$ for all $f \in F$ is learnable with respect to a distribution P if $N(\epsilon, F, d_P) < \infty$ for every $\epsilon > 0$. Furthermore, a sample size*

$$m(\epsilon, \delta) \geq \frac{2M^2}{\epsilon^2} \ln \frac{2N(\epsilon, F, d_P)}{\delta}$$

is sufficient for ϵ, δ learnability.

This theorem is simply a special case of Theorem 1 from [23]. Note that in the above result, it is actually not necessary for every $f \in F$ to be bounded. We really only need the functions in the ϵ-cover to be bounded, and the bound can depend on ϵ. Specifically, in the sample size bound in the above theorem M could be replaced by $M(\epsilon/2)$ where $M(\epsilon/2)$ is a bound on all the functions in an $\epsilon/2$-cover.

In the case distribution-free learning, a quantity which generalizes the notion of VC dimension provides a sufficient condition for learnability. Following [23] we refer to this quantity as pseudo dimension.

Definition 8 (Pseudo Dimension) *Let F be a collection of functions from a set X to $\mathbf{R}$. For any set of points $\overline{x} = (x_1, \ldots, x_d)$ from X, let $F_{|\overline{x}} = \{(f(x_1), \ldots, f(x_d)) : f \in F\}$. $F_{|\overline{x}}$ is a set of points in $\mathbf{R}^d$. If there is some translation of $F_{|\overline{x}}$ which intersects all of the 2^d orthants of $\mathbf{R}^d$*

then $\bar{x}$ is said to be shattered *by F. Following terminology from [23], the* pseudo dimension *of F, which we denote* $\dim(F)$, *is the largest integer d such that there exists a set of d points in X that is shattered by F. If no such largest integer exists then* $\dim(F)$ *is infinite.*

The idea of this definition is that, given a set of samples $(x_1, a_1), \ldots, (x_n, a_n)$, each $f \in F$ separates the samples into two sets — those that lie above f and those that lie below f (ignoring the case when f passes exactly through one or more of the points). If all 2^n separations of the n points can be formed using functions from F then the n points are shattered. Thus, this definition is very similar to the previous definition of VC dimension (and in fact reduces to the previous definition in the case of 0,1-valued functions). As a simple example of pseudo dimension, it can be easily shown that if $X = \mathbf{R}$ and F is the set of all functions of the form $f(x) = ax + b$ then $\dim(F) = 2$.

The following theorem gives a sufficient condition and sample size bounds for distribution-free learnability. Again, this theorem is also a special case of a result from [23] (i.e., Corollary 2 from [23]).

Theorem 9 *A collection of functions F with $0 \leq f \leq M$ for every $f \in F$ is distribution-free learnable if* $\dim(F) = d$ *for some $1 \leq d < \infty$. Furthermore, a sample size*

$$m(\epsilon, \delta) \geq \frac{64M^2}{\epsilon^2}\left(2d \ln \frac{16eM}{\epsilon} + \ln \frac{8}{\delta}\right)$$

is sufficient for ϵ, δ distribution-free learnability.

Note that in Theorems 8 and 9 only sufficient conditions for learnability were given. Finite metric entropy and finite pseudo dimension are not in general necessary for fixed distribution and distribution-free learnability respectively. For example, take $X = [0, 1]$ and consider a fixed distribution which has a small positive mass at zero and is otherwise uniform over $[0, 1]$. We can consider a class of functions F indexed by a bounded real (or countable) parameter α which has infinite metric entropy over the interval $(0, 1]$, but such that $f_\alpha(0) = \alpha$. Thus, the value of the function at $x = 0$ completely encodes the function. Since there is positive mass at $x = 0$, F is clearly learnable, but by construction has infinite metric entropy. However, note that noisy samples would create problems with this example. For distribution free learning, we can consider a class of functions with infinite pseudo dimension but such that for any ϵ only a finite number of functions are ϵ-far from the zero function for every distribution. This could be done, for example, by taking a countable set of bounded functions with infinite pseudo dimension and weighting the i-th function by $1/i$. Then for ϵ, δ learning, we need only decide among a finite number of functions which can be done to confidence δ using a uniformly bounded number of samples.

Since the present formulation is a special case of the formulation in [23], the comments about sufficient but not necessary conditions apply to the more general formulation of [23] as well. One necessary condition for learnability was obtained by Kearns and Schapire [26] in terms of a variation of pseudo dimension in which the shattering is required to take place with separation greater than some fixed constant. However, as far as we know, the problem of finding a single condition which is both necessary and sufficient in the case of either fixed distribution or distribution-free learnability is open. Also, as with the basic PAC model, the problem of characterizing learnability for a general class of distributions is also open.

6 Generalized Samples

In the previous section, the information received by the learner consisted of random samples of some unknown function. In [29, 32], an extension was considered in which the learner may receive information from much more general types of samples, referred to as generalized samples. A generalized sample is essentially a functional assigning a real number to each concept, where the number assigned may not necessarily be the value of the unknown concept at a point, but could be some other attribute of the unknown concept (e.g., the integral over a region, or the derivative at a given point, etc.).

As in the previous section, let X be the original instance space, and let the concept class F be a collection of real valued functions on X. In the usual model as described in the previous sections, the information the learner gets are samples $(x, f(x))$ where $x \in X$ and where $f \in F$ is the target concept. This can be viewed as obtaining a functional δ_x and applying this functional to the target concept f to obtain the sample $(\delta_x, \delta_x(f)) = (\delta_x, f(x))$. The functional in this case simply evaluates f at the point x, and is chosen randomly from the class of all such "impulse" functionals. Instead, we now assume we get generalized samples in the sense that we obtain a more general functional $\tilde{x}$, which is some mapping from F to $\mathbf{R}$. The observed labeled sample is then $(\tilde{x}, \tilde{x}(f))$ consisting of the functional and the real number obtained by applying this functional to the target concept f. We assume the functional $\tilde{x}$ is chosen randomly from some collection of functionals $\tilde{X}$. Thus, $\tilde{X}$ is the instance space for the generalized samples, and the distribution P is a probability measure on $\tilde{X}$. Let S_F denote the set of labeled m-samples for each $m \geq 1$, for each $\tilde{x} \in \tilde{X}$, and each $f \in F$. The natural error criterion for generalized samples is

$$d_P(f_1, f_2) = E_P|\tilde{x}(f_1) - \tilde{x}(f_2)|$$

which is simply the average absolute difference of the real numbers produced by generalized samples on the two concepts.

Definition 9 (Learning From Generalized Samples) *Let $\mathcal{P}$ be a fixed and known collection of probability measures. Let F be a collection of functions from the instance space X into $\mathbf{R}$, and let $\tilde{X}$ be the instance space of generalized samples for F. F is said to be* learnable with respect to $\mathcal{P}$ from the generalized samples $\tilde{X}$ *if there is a mapping $A : S_F \to F$ for producing a hypothesis h from a set of labeled samples such that for every $\epsilon, \delta > 0$ there is a $0 < m = m(\epsilon, \delta) < \infty$ such that for every probability measure $P \in \mathcal{P}$ and every $f \in F$, if h is the hypothesis produced from a labeled m-sample drawn according to P^m then the probability that $d_P(f, h) < \epsilon$ is greater than $1 - \delta$.*

Learning with generalized samples can be easily transformed into an equivalent problem of PAC learning from standard samples. The concept class F on X corresponds naturally to a concept class $\tilde{F}$ on $\tilde{X}$ as follows. For a fixed $f \in F$, each functional $\tilde{x} \in \tilde{X}$ produces a real number when applied to f. Therefore, f induces a real valued function on $\tilde{X}$ in a natural way. The real valued function on $\tilde{X}$ induced by f will be denoted by $\tilde{f}$, and is defined by

$$\tilde{f}(\tilde{x}) = \tilde{x}(f)$$

The concept class $\tilde{F}$ is the collection of all functions on $\tilde{X}$ obtained in this way as f ranges through F.

We are now in the standard PAC framework with instance space $\tilde{X}$, concept class $\tilde{F}$, and distribution P on $\tilde{X}$. Hence, as usual, P induces a learning criterion or metric (actually only a pseudo-metric in general) on $\tilde{F}$, and as a result of the correspondence between F and $\tilde{F}$, this metric is the equivalent to the pseudo-metric d_P induced by P on F. This metric will be denoted by d_P over both F and $\tilde{F}$, and is given by

$$d_P(\tilde{f}_1, \tilde{f}_2) \quad = \quad E|\tilde{f}_1 - \tilde{f}_2| = E|\tilde{x}(f_1) - \tilde{x}(f_2)| = d_P(f_1, f_2)$$

Distribution-free and fixed distribution learnability are defined in the usual way for $\tilde{X}$ and $\tilde{F}$. Thus, the pseudo dimension and metric entropy of $\tilde{F}$ characterize the learnability of $\tilde{F}$ in the distribution-free and fixed distribution cases respectively. These same quantities for $\tilde{F}$ then also characterize the learnability of F with respect to d_P. The following results for fixed distribution and distribution-free learning from generalized samples were shown in [29, 32] using results from [23].

Theorem 10 *F is learnable from generalized samples (or equivalently, $\tilde{F}$ is learnable) with respect to a distribution P if for each $\epsilon > 0$ there is a finite ϵ-cover $\tilde{F}^{(\epsilon)}$ for $\tilde{F}$ (with respect to d_P) such that $0 \leq f_i \leq M(\epsilon)$ for each $f_i \in \tilde{F}^{(\epsilon)}$. Furthermore, a sample size*

$$m(\epsilon, \delta) \geq \frac{2M^2(\epsilon/2)}{\epsilon^2} \ln \frac{2|\tilde{F}^{(\epsilon/2)}|}{\delta}$$

is sufficient for ϵ, δ learnability.

Theorem 11 *F is distribution-free learnable from generalized samples (or equivalently, $\tilde{F}$ is distribution-free learnable) if for some $M < \infty$ we have $0 \leq \tilde{f} \leq M$ for every $\tilde{f} \in \tilde{F}$ and if $\dim(\tilde{F}) = d$ for some $1 \leq d < \infty$. Furthermore, a sample size*

$$m(\epsilon, \delta) \geq \frac{64M^2}{\epsilon^2} \left(2d \ln \frac{16eM}{\epsilon} + \ln \frac{8}{\delta} \right)$$

is sufficient for ϵ, δ distribution-free learnability.

Note that the metric entropy of $\tilde{F}$ is identical to the metric entropy of F (since both are with respect to d_P), so that the metric entropy of F characterizes learnability for a fixed distribution as well. However, the pseudo dimension of F with respect to X does *not* characterize distribution-free learnability. This quantity can be very different from the pseudo dimension of $\tilde{F}$ with respect to $\tilde{X}$. As far as we know, very few computations of metric entropy and/or pseudo dimension have been carried out for generalized samples. We expect that as with regular samples, such computations (or bounds) can be carried out in a number of cases.

As mentioned above, for simplicity we have defined the concepts to be real valued functions, have chosen the generalized samples to return real values, and have selected a particular form for the learning criterion or metric d_P. The ideas can easily be formulated in the much more general framework considered by Haussler [23]. Specifically, one could take F to be a family of functions with domain X and range A. The generalized samples $\tilde{X}$ would be a collection of mappings from F to $\tilde{A}$. A family of functions $\tilde{F}$ mapping $\tilde{X}$ to $\tilde{A}$ would be obtained from F by assigning to each $f \in F$ an $\tilde{f} \in \tilde{F}$ defined by $\tilde{f}(\tilde{x}) = \tilde{x}(f)$. As in [23], the distributions would be defined on $\tilde{X} \times \tilde{Y}$, a loss function L would be defined on $\tilde{Y} \times \tilde{A}$, and for each $\tilde{f} \in \tilde{F}$ the error of the hypothesis $\tilde{f}$ with respect to a distribution would be $EL(\tilde{f}(\tilde{x}), \tilde{y})$ where the expectation is over the distribution on $(\tilde{x}, \tilde{y})$.

An interesting geometric example of learning from generalized samples was considered in [32]. Namely, take X to be the unit square in the plane, and consider concept classes which are collections of curves contained in X. For example, one simple concept class is the set of straight line segments contained in X. A much more general concept class considered is the set of curves in X with bounded length and bounded turn (a notion which generalizes total absolute curvature to not necessarily differentiable curves – e.g., see [1]). The samples observed by the learner consist of randomly drawn straight lines labeled as to the number of intersections the random line makes with the target concept (i.e., the unknown curve). Learnability with respect to a fixed distribution was considered, where the distribution is the uniform distribution on the set of lines intersecting X. A learnability result was obtained by providing metric entropy bounds for the class of curves under considered.

The example of learning a curve is closely related to a classical result from stochastic geometry which states that the expected number of intersections a random line makes with an arbitrary rectifiable curve is proportional to the length of the curve (e.g., see [48]). This result suggests that the length of a curve can be estimated (or "learned") from a set of generalized samples. In fact, this idea has been studied, although primarily from the point of view of deterministic sampling [52, 39]. The learnability result makes the much stronger statement that for certain classes of curves, from just knowing the number of intersections with a set of random lines, the curve itself can be learned (from which the length can then be estimated). Also, for these classes of curves, the learning result guarantees uniform convergence of empirical estimates of length to true length, which does not follow directly from the stochastic geometry result.

Although learning with generalized samples is essentially just a transformation to a standard learning problem, the variety available in choosing this transformation (i.e., the form of the generalized samples) should allow the learning framework and results to be applied to a broad range of problems. In particular, this framework could be applied to a number of problems in signal/image processing, geometric reconstruction, stereology, etc., to provide learnability results and sample size bounds under a PAC criterion.

For example, the generalized samples could consist of drawing certain random sets and returning the integral of the concept over these sets. Other possibilities might be to return weighted integrals of the concept where the weighting function is selected randomly from a suitable set (e.g., an orthonormal basis), or to sample derivatives of the concept at random points. One interesting application would be to problems in tomographic reconstruction. In these problems, one is interested in reconstructing a function from a set of projections of the function onto lower dimensional subspaces. One could have the generalized samples consist of drawing random lines labeled according to the integral of the unknown function along the line. This would correspond to a problem in tomographic reconstruction with random ray sampling. Alternatively, one could consider an application to tomography where the generalized samples consist of entire projections. This would be more in line with standard problems in tomography, but with the directions of the projections being chosen randomly.

For more geometric problems in which the concepts are subsets of X, some interesting generalized samples might be to draw random (parameterized) subsets (e.g., disks, lines, or other parameterized curves) of X labeled as to whether or not the random set intersects or is contained in the target concept. Other possibilities might be to label the random set as to the number of intersections (or length, area, or volume of the intersection, as appropriate) with the unknown concept. One interesting application is the reconstruction of a convex set from various types data as considered in [30]. This is related to other work on convex set reconstruction (e.g., see [24, 50, 35, 43]).

For such applications, the main problem is to perform the metric entropy computation which in general can be quite difficult. Another problem which can be difficult is interpreting the learning criterion on the original space induced by the distribution on the generalized samples. The induced metric is a natural one given the type of information available, but it may be difficult to understand the properties it endows on the original concept class.

In addition to applications, there are a number of open problems dealing directly with learning from generalized samples framework. Essentially every open problem for learning functions is open for learning with generalized samples as well. For example, as with Theorems 8 and 9, Theorems 10 and 11 provide sufficient but not necessary conditions for learnability. Finding necessary and sufficient conditions for learnability with respect to a fixed distribution, all distributions, and a class of distributions are all open.

7 Discussion

We considered several extensions of the basic PAC model focusing on the information complexity of learning. There has been a tremendous amount of work done on extensions and applications of the basic PAC model. Here we briefly mention some other work concerned with the information complexity of learning.

An important extension which we did not discuss above is nonuniform learning. That is, in all the models we discussed the sample size required for learning was required to be uniformly bounded over all concepts and all probability measures. A natural extension to consider is to allow the number of samples to depend on the concept and/or probability measure. Some work has been done along these lines. In fact, this idea appeared in Vapnik's early work [57], for which he used the term "structural risk minimization". Valiant's formulation also contained this idea. Blumer et al. [10, 11], Linial et al. [36], Benedek and Itai [8], and Ben-David et al. [5] all considered this idea as extensions of Valiant's framework as well. Nonuniform learning has been primarily considered for either the basic PAC model or for learning functions. It may be interesting to consider nonuniform learning for some extensions such as learning over a class of distributions or learning with queries. For example, nonuniform active learning is closely related to universal coding ideas [17, 44, 45] from information theory.

Some extensions to the basic PAC model are covered by the formulation of [23] (e.g., the notion of learning probabilistic concepts as studied in [26]). However, there are other interesting extensions we have not discussed which do not directly fall under the paradigm in [23] (e.g., nonuniform learning mentioned above, learning by distances [6], simultaneous learning [9], learning with a helpful teacher [47]). Finally, in addition to studying extensions to the PAC model, an interesting and important direction is to study connections of PAC learning with other learning and estimation paradigms [20, 22, 29, 33, 37, 38, 45, 58].

References

[1] Alexandrov, A.D. and Yu.G. Reshetnyak, *General Theory of Irregular Curves*, Mathematics and Its Applications (Soviet Series) Vol. 29, Kluwer Academic Publishers, 1989.

[2] Amsterdam, J., "Extending the Valiant learning model," *Proc. 5th Int. Conf. on Machine Learning*, pp. 381-394, 1988.

[3] Angluin, D. and P. Laird, "Learning from noisy examples," *Machine Learning*, Vol. 2, pp.343-370, 1988.

[4] Angluin, D., "Queries and Concept Learning," *Machine Learning*, Vol. 2, pp.319-342, 1988.

[5] Ben-David, S., G.M. Benedek, Y. Mansour, "A parameterization scheme for classifying models of learnability," *Proc. Second Annual Workshop on Computational Learning Theory*, pp. 285-302, 1989.

[6] Ben-David, S., A. Itai, E. Kushilevitz, "Learning by distances," *Proc. of Third Annual Workshop on Computational Learning Theory*, pp. 232-245, 1990.

[7] Benedek, G.M. and A. Itai, "Learnability with respect to fixed distributions," *Theoretical Computer Science*, Vol. 86(2), pp. 377-390, 1991.

[8] Benedek, G.M. and A. Itai, "Nonuniform learnability," *ICALP*, pp. 82-92, 1988.

[9] Buescher, K. and P.R. Kumar, "Simultaneous learning and estimation for classes of probabilities," *Proc. Fourth Annual Workshop on Computational Learning Theory*, Santa Cruz, CA, Aug 1991.

[10] Blumer, A., A. Ehrenfeucht, D. Haussler, M. Warmuth, "Occam's razor," *Info. Proc. Let.*, Vol. 24, pp. 377-380, 1987.

[11] Blumer, A., A. Ehrenfeucht, D. Haussler, M. Warmuth, "Learnability and the Vapnik-Chervonenkis dimension," *J. ACM*, Vol. 6, No. 4, pp. 929-965, 1989.

[12] Dudley, R.M., "Central limit theorems for empirical measures," *Ann. Probability*, Vol. 6, No. 6, pp. 899-929, 1978.

[13] Dudley, R.M., "Metric entropy of some classes of sets with differentiable boundaries," *J. Approx. Theory*, Vol. 10, No. 3, pp. 227-236, 1974.

[14] Dudley, R.M., S.R. Kulkarni, T.J. Richardson, and O. Zeitouni, "A metric entropy bound is not sufficient for learnability," *IEEE Trans. Information Theory*, Vol. 40, No. 3, pp. 883-885, 1994.

[15] Ehrenfeucht, A., D. Haussler, M. Kearns, and L. Valiant, "A general lower bound on the number of examples needed for learning," *Information and Computation*, Vol. 82, No. 3, pp. 247-251, 1989.

[16] Eisenberg, B. and R.L. Rivest, "On the Sample Complexity of Pac-Learning Using Random and Chosen Examples," *Proc. Third Annual Workshop on Computational Learning Theory*, pp. 154-162, 1990.

[17] Elias, P., "Universal codeword sets and representations of the integers," *IEEE Trans. on Info. Theory*, Vol. IT-21, No. 2, pp. 194-203, 1975.

[18] Gallager, R.G., *Information Theory and Reliable Communication*, Wiley & Sons, 1968.

[19] Gasarch, W.I and C.H. Smith, "Learning via queries," *Proc. of the 29th IEEE Symp. on Foundations of Computer Science*, 1988.

[20] Gold, I.M, "Language identification in the limit," *Information and Control*, Vol. 10, pp. 447-474, 1967.

[21] Haussler, D., M. Kearns, N. Littlestone, M.K. Warmuth, "Equivalence of models for polynomial learnability," *Proc. First Workshop on Computational Learning Theory*, pp. 42-55, 1988.

[22] Haussler, D., M. Kearns, R. Schapire, "Bounds on the sample complexity of Bayesian learning using information theory and the VC dimension," *Proc. Fourth Annual Workshop on Computational Learning Theory*, Santa Cruz, CA, Aug 1991.

[23] Haussler, D., "Decision theoretic generalizations of the PAC model for neural net and other learning applications," *Information and Computation*, Vol. 100, pp. 78-150, 1992.

[24] Karl, W.C., "Reconstructing objects from projections," Ph.D. Thesis, Dept. of EECS, Massachusetts Institute of Technology, February, 1991.

[25] Kearns, M. and M. Li, "Learning in the presence of malicious errors," *Proc. 20th ACM Symp. on Theory of Comp.*, Chicago, Illinois, pp. 267-279, 1988.

[26] Kearns, M.J. and R.E. Schapire, "Efficient distribution-free learning of probabilistic concepts," *Proc. of the 31th IEEE Symp. on Foundations of Computer Science*, 1990.

[27] Kinber, E.B., "Some problems of learning with an oracle," *Proc. Third Workshop on Computational Learning Theory*, pp. 178-186, 1990.

[28] Kolmogorov, A.N. and V.M. Tihomirov, "ϵ-Entropy and ϵ-capacity of sets in functional spaces," *Amer. Math. Soc. Transl.*, Vol. 17, pp. 277-364, 1961.

[29] Kulkarni, S.R., "Problems of computational and information complexity in machine vision and learning," Ph.D. thesis, Dept. of Electrical Engineering and Computer Science, M.I.T., June, 1991.

[30] Kulkarni, S.R., "Applications of PAC learning to problems in geometric reconstruction," *Proc. 27th Annual Conf. on Info. Sciences and Systems*, Johns Hopkins University, March, 1993.

[31] Kulkarni, S.R., S.K. Mitter, J.N. Tsitsiklis, "Active learning using arbitrary binary valued queries," *Machine Learning*, Vol. 11, pp. 23-35, 1993.

[32] Kulkarni, S.R., S.K. Mitter, J.N. Tsitsiklis, and O. Zeitouni, "PAC learning with generalized samples and an application to stochastic geometry," *IEEE Trans. Pattern Analysis and Machine Intelligence*, Vol. 15, No. 9, pp. 1-10, 1993.

[33] Kulkarni, S.R. and D.N.C. Tse, "A paradigm for class identification problems," *IEEE Trans. Information Theory*, Vol. 40, No. 3, pp. 696-705, 1994.

[34] Kulkarni, S.R. and M. Vidyasagar, "Learning decision rules for pattern classification under a family of probability measures," submitted to *IEEE Trans. Information Theory.*

[35] Lele, A.S., S.R. Kulkarni, and A.S. Willsky, "Convex set estimation from support line measurements and applications to target reconstruction from laser radar data," *J. Optical Soc. of Amer. A*, Vol. 9, No. 10, 1992.

[36] Linial, N., Y. Mansour, R.L. Rivest, "Results on learnability and the Vapnik-Chervonenkis dimension," *Proc. First Workshop on Computational Learning Theory*, pp. 56-68, 1988.

[37] Littlestone, N., "Mistake bounds and logarithmic linear-threshold learning," Ph.D. thesis, U.C. Santa Cruz, March, 1989.

[38] Littlestone, N., "Learning when irrelevant attributes abound: A new linear-threshold algorithm," *Machine Learning*, Vol. 2, pp. 285-318, 1988.

[39] Moran, P.A.P., "Measuring the length of a curve," *Biometrika*, Vol. 53, pp. 359-364, 1966.

[40] Natarajan, B.K., "Probably approximate learning over classes of distributions," Carnegie-Mellon Univ., unpublished manuscript, 1989.

[41] Natarajan, B.K. and P.T. Tadepalli, "Two new frameworks for learning," *Proc. 5th Int. Conf. on Machine Learning*, pp. 402-415, 1988.

[42] Pollard, D., *Convergence of Stochastic Processes*, Springer-Verlag, 1984.

[43] Prince, J.L. and A.S. Willsky, "Estimating convex sets from noisy support line measurements," *IEEE Trans. PAMI*, Vol. 12, pp. 377-389, 1990.

[44] Rissanen, J., "A universal prior for the integers and estimation by minimum description length," *Annals of Statistics*, Vol. 11, No. 2, pp.416-431, 1983.

[45] Rissanen, J., *Stochastic Complexity in Statistical Inquiry*, Series in Computer Science Vol. 15, World Scientific, 1989.

[46] Rivest, R.L., A.R. Meyer, D.J. Kleitman, K. Winklmann, and J. Spencer, "Coping with errors in binary search procedures," *J. of Computer and System Sciences*, Vol. 20, pp. 396-404, 1980.

[47] Salzberg, S., A. Delcher, D. Heath, and S. Kasif, "Learning with a helpful teacher," Technical Report, Dept. of Computer Science, Johns Hopkins University, 1990.

[48] Santalo, L.A., *Integral Geometry and Geometric Probability.* Volume 1 of *Encyclopedia of Mathematics and its Applications,* Addison-Wesley, Reading, MA, 1976.

[49] Schapire, R., "The strength of weak learnability," *Machine Learning*, Vol. 5, pp.197-227, 1990.

[50] Skiena, S.S., "Geometric probing," Ph.D. thesis, Dept. of Computer Science, Univ. of Illinois at Urbana-Champaign, (report no. UIUCDCS-R-88-1425), April, 1988.

[51] Sloan, R., "Types of noise in data for concept learning," *Proc. First Workshop on Computational Learning Theory*, pp. 91-96, 1988.

[52] Steinhaus, H., "Length, shape, and area," *Colloquium Mathematicum*, Vol. 3, pp. 1-13, 1954.

[53] Tikhomirov, V.M., "Kolmogorov's work on ϵ-entropy of functional classes and the superposition of functions," *Russian Math. Surveys*, Vol. k8, pp. 51-75, 1963.

[54] Valiant, L.G., "A theory of the learnable," *Comm. ACM*, Vol. 27, No. 11, pp. 1134-1142, 1984.

[55] Vapnik, V.N. and A.Ya. Chervonenkis, "On the uniform convergence of relative frequencies to their probabilities," *Theory of Prob. and its Appl.*, Vol. 16, No. 2, pp. 264-280, 1971.

[56] Vapnik, V.N. and A.Ya. Chervonenkis, "Necessary and and sufficient conditions for the uniform convergence of means to their expectations," *Theory of Prob. and its Appl.*, Vol. 26, No. 3, pp. 532-553, 1981.

[57] Vapnik, V.N., *Estimation of Dependences Based on Empirical Data*, Springer-Verlag, 1982.

[58] Yamanishi, K., "A learning criterion for stochastic rules," *Proc. Third Workshop on Computational Learning Theory*, pp. 67-81, 1990.

GEOMETRY

Finite Point Sets and Oriented Matroids

Combinatorics in Geometry

Jürgen Bokowski
Department of Mathematics
University of Darmstadt
Schloßgartenstraße 7
D-64289 Darmstadt, Germany
e-mail: bokowski@mathematik.th-darmstadt.de

Abstract. Many algorithms in computational geometry deal with finite point sets represented by matrices over the reals. The set of oriented matroids can be considered as a set of topological invariants of matrices over the reals together with a natural set of such topological objects making the set complete. One main advantage of these topological invariants is its possible representation as signed vectors allowing computational investigations easily. Various systems of axioms exist for oriented matroids. We try to illustrate basic ideas in the theory of oriented matroids by using various interesting example classes.

Example. Consider the ordered set of vertices $\{p_0, p_1, \ldots, p_{11}\}$ of the regular icosahedron. For each vector $v_i := p_i - p_0$, $i = 1, 2, \ldots, 11$, we have a corresponding oriented plane E_i through p_0 orthogonal to v_i. The ordered set of oriented planes $\{E_1, \ldots, E_{11}\}$ induces a cell decomposition on the unit sphere around the vertex p_0. The topological information of this cell decomposition is *the oriented matroid $\mathcal{M}$ of the set of vectors* $\{v_1, \ldots, v_{11}\}$. Fig. 1 on the next page shows a corresponding model of the cell decomposition without marked orientations.

1. Introduction

In computational geometry we study algorithms involving finite point sets represented by matrices over the reals. The main objective of this article is showing that topological invariants of such point sets or topological invariants of matrices over the reals (oriented matroids) can be of much help in computational geometry. One main advantage of these topological invariants is the possibility of storing the invariant information as a signed vector. It allows computational investigations easily, especially, for generating geometric objects with prescribed combinatorial properties. We consider the row vectors of a matrix as representing normal vectors of oriented hyperplanes passing through the origin. The oriented hyperplanes define a cell decomposition of the unit sphere around the origin. This is a topological information of our given matrix we are interested in. One closed halfsphere with its cell decomposition carrying the information about the orientations already defines the complete topological invariant. We consider next some examples in order to explain this concept.

Fig. 1. Model of the vertex figure of the icosahedron.

1.1. Vertex figures of Platonic solids

We consider the set of vertices $\{p_0, p_1, \ldots, p_{11}\}$ of the regular icosahedron. Each difference vector $v_i := p_i - p_0,\ i = 1, 2, \ldots, 11$, defines a corresponding oriented plane E_i through p_0 orthogonal to v_i. The set of oriented planes $\{E_1, \ldots, E_{11}\}$ in turn induces a cell decomposition on the unit sphere around the vertex p_0. The topological information of this cell decomposition is *the oriented matroid $\mathcal{M}$ of the set of vectors* $\{v_1, \ldots, v_{11}\}$. The picture in Fig. 1 shows a corresponding model of the cell decomposition without marked orientations. This topological information can also be seen from a projection of one closed half of the 2 sphere. This is what we see in the drawings in Fig. 2. The left part corresponds to what we see in the actual picture and the right part corresponds to its mirror image. We come back to this example when studying the Goodman–Pollack bad pentagon.

Fig. 2. Vertex figure of the icosahedron (2 versions of the same cell decomposition).

We next consider all 19 rays starting from the vertex v of a regular dodecahedron and pointing to its remaining vertices, and we consider them as defining oriented planes orthogonal to these rays. These oriented planes induce on the unit sphere around this vertex v a cell decomposition, the topological information of which (its oriented matroid) is seen when projecting one bounded half of this sphere down onto a plane, see the colored Fig. 14.

In the remaining cases of Platonic solids the vertex figures are simpler. The case of the cube in all dimensions was studied in [14]. For the 3-cube and the octahedron we have the corresponding topological information in Fig. 3. The orientations have not been indicated.

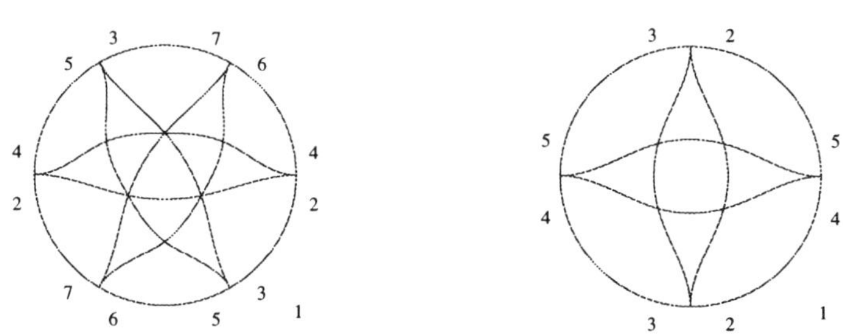

Fig. 3. Vertex figure of the cube (left) and vertex figure of the octahedron (right).

1.2. A projection of the 600-cell onto a plane

This example occured during investigating a problem of P. McMullen posed in his Ph.D.-dissertation: there is a 3-sphere obtainable from the regular 600-cell for which it is unknown whether it can be obtained as the boundary of a convex 4-polytope. This 3-sphere S has like the 600-cell 120 vertices and its polar dual coincides with S. Assuming that S can be obtained as the boundary of a 4-polytope, the colored Fig. 14 would show a two-element contraction of this questionable polytope.

For understanding this example, we might start directly by presenting the corresponding matrix of rank 3. But we prefer to show its geometric background, too. The regular 600-cell in dimension 4 can be written in homogeneous coordinates: a 120×5-matrix. We take a vertex of the 600-cell as the center for a central projection onto a 3 dimensional hyperplane. From this image we start once more a central projection onto a 2 dimensional plane, the new center being the image of the opposite vertex of the first chosen center (with respect to the 600-cell). The result is a point set in the plane which can be described with homogeneous coordinates as a matrix with 3 columns. This is our matrix, the topological properties of which we want to understand. We interpret the rows of our matrix as outer normal vectors of planes passing through the origin, we consider the induced cell decomposition of the 2-sphere around the origin, and we project down the corresponding topological cell decomposition of the 2-sphere onto the plane, one half of it being sufficient. The corresponding topological picture is shown in the colored Fig. 14. We have obtained only a homeomorphic image of our cell decomposition, but since we are interested only in the topological structure of the cell decomposition, this shows us everything. We can mark the orientations of our former planes in this picture, and we can label the curves according to the labels of the vertices of the 600-cell. With this example (after picking orientations for all curves in the circle), the reader gets a first impression what an oriented matroid looks like: the picture represents the equivalence class of homeomorphic images of the cell decomposition of the 2-sphere as explained above: the oriented matroid of the matrix of rank 3.

We can also see the example this way: Consider all rays r_i, $i = 1, \ldots, 599$ starting from the vertex v of the regular 600-cell and pointing to its remaining vertices v_i, and consider them as defining oriented hyperplanes H_i orthogonal to these rays r_i. Let r_1 be the ray pointing to the opposite point of v and consider a hyperplane H through v together with one of its defining closed halfspaces H^+. The oriented hyperplanes H_i induce on the unit sphere $S^3 := \{x \mid dist(x,v) = 1\}$ around this vertex v a cell decomposition, the topological information of which can be seen when projecting orthogonally one bounded half $S^3 \cap H^+$ of S^3 down onto the 3-dimensional space H. This provides a cell decomposition of the unit ball on which antipodal points have to be identified. This topological cell decomposition (information about the orientations is kept, information about the boundaries of the cells belonging to the same topological hyperplane is kept as well) can be considered as the oriented matroid of the vertex figure of the 600-cell. What we see in Fig. 14 is the topological information of the image of $S^3 \cap H^+ \cap H_1$ for a suitable hyperplane H.

1.3. The topological concept of oriented matroids generalizes that of matrices.

We proceed by providing some main ideas of the theory of oriented matroids. Consider now having a point configuration of n points in a real vectorspace of fixed dimension d.

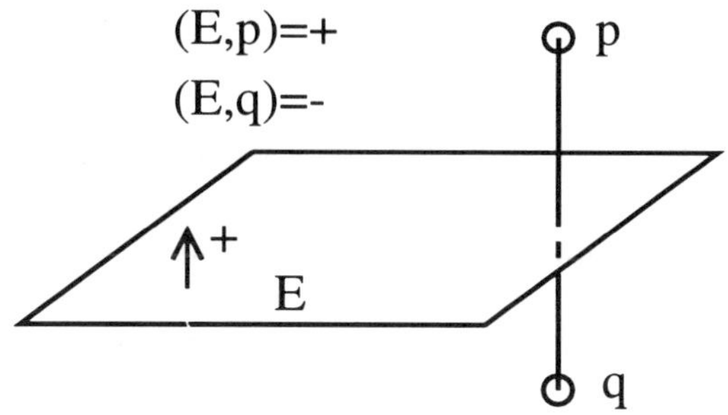

Fig. 4. Finite point set. Positions of points with respect to oriented planes defined as affine hulls of other subsets of these points.

We assume all $(d-1)$-dimensional planes, generated as affine hulls of subsets of these points, to be oriented, and we store the combinatorial information which point lies on which side relatively to these oriented $(d-1)$-dimensional planes. The resulting combinatorial information of the point configuration turns out to be equivalent to the foregoing concept, ie it is invariant under topological transformations of the space, and it can be viewed as the (in this case: *realizable*) oriented matroid of the point configuration.

Let us repeat the concept we explained before for Fig. 14 in the 3-dimensional affin case. A 3-dimensional point configuration $\{1, 2, \ldots, i, \ldots, n\}$ in the hyperplane $\{(w, x, y, z) \mid w = 1\}$ leads to rays r_i from the origin $(0,0,0,0)$ to the points i which have orthogonal (oriented) 3-dimensional hyperplanes H_i containing the origin. The intersection of H_i with the 3-sphere $\{(w, x, y, z) \mid w^2+x^2+y^2+z^2 = 1\}$ is an oriented 2-sphere. Any two such 2-spheres intersect in an (oriented) 1-sphere which has intersection points (so called cocircuits) with all remaining 2-spheres. This sequence of cocircuits along a 1-sphere gives back the signs of all 4×4-determinants of 4 former points with coordinates $(1, x, y, z)$.
There are various systems of axioms to describe this structure, all of which generalize the realizable point configurations. By gaining a precise (and from many viewpoints very natural) definition of generalized point configurations, we lose the familiar realized ones as representatives in the general case.
A general method in solving problems is reducing the complexity and studying smaller examples first, before tackling the more general question again. It is also possible that by generalizing and embedding the problem in a more general context, we gain essential insight, and we have sufficient skill to solve the original problem. Think of the fundamental theorem of algebra, concerning the real zeros of a polynomial in one variable. The better insight for properties of real zeros is gained by embedding the solution set into the closed field of complex numbers:

There are as many zeros in the field of complex numbers as the degree of the polynomial indicates.

Real solutions $\Longleftrightarrow$ complex solutions.

A similar fact occurs when solving problems of point configurations where the positions of points with respect to oriented planes play an essential role. Oriented matroids form a closed set of topological invariants that we have introduced above.

Signs of sub-determinants of matrices $\Longleftrightarrow$ signs of bases of oriented matroids.

Problems in which these topological invariants play a key role occur for instance in computational synthetic geometry [22]. The objective of computational synthetic geometry is finding geometrical objects or proving their non-existence when only partial, especially combinatorial data of the object is given. The convexity property of a polyhedron presented by its vertices says that no vertex lies on the outer normal side of any supporting hyperplane. For problems involving such assertions, the theory of oriented matroids has turned out to be an important general tool. Once an overview from the more general concept of oriented matroids has been accomplished, we can look back to the problem of point configurations, we started with.

Just as complex numbers help to understand the structure of solution sets of polynomials in the real case, oriented matroids help finding point sets with prescribed properties.

In the same manner as there is a resistance to study complex numbers when faced with them the first time, there seems to be a resistance to accept the structure of oriented matroids as a natural one, when studying point sets in real space with their topological (or combinatorial) properties.

Within this article, we try to brake such a resistance by introducing some basic concepts of oriented matroids without frightening the reader with too much of a technical language. The objective is to motivate and to prepare the reader for further studies in this field. Whenever possible, we try to use samples of oriented matroids to understand essential properties of them. We understand the word *sample* as a piece or item that shows the quality of the whole, and thinking in the spirit of this volume, we quote Haim Shvaytser's first sentence in his article *Learnable and nonlearnable visual concepts*: "Often, learning from examples appears to be easy for humans, but difficult for artificial machines".

The examples we are going to study have been used by the author for motivating students to enter the field or to learn basic geometric features and properties of otherwise more difficult and more abstract objects. At the same time, these objects are linked, or were linked, to open research problems.

Oriented matroids are closely related to vectorspaces. The close relation to such a fundamental concept in mathematics might be the reason for why so many mathematicians independently reinvented this structure in very often completely different areas of mathematics and its applications, just as there are so many different applications of linear algebra. Applications to convex polytopes and general polyhedral sets have been already mentioned, we add intersection properties of line-segments with triangles as they occur in computer graphics or within robotics, duality concepts as part of linear programming, the classification of molecules in chemistry.

1.4. Line segments piercing triangles

Engineers use to work with points as vectors with coordinates rather than with abstract objects. Especially, checking of whether there exists a point of intersection between a triangle and a line segment would mean for them automatically calculating the point of intersection and therefore using metrical data. Although, these computations are done very fast, this might not be the appropriate tool, especially, when there are many of these checkings involved. Moreover, there are many problems in which the coordinates are not available whereas these checkings can still be done.

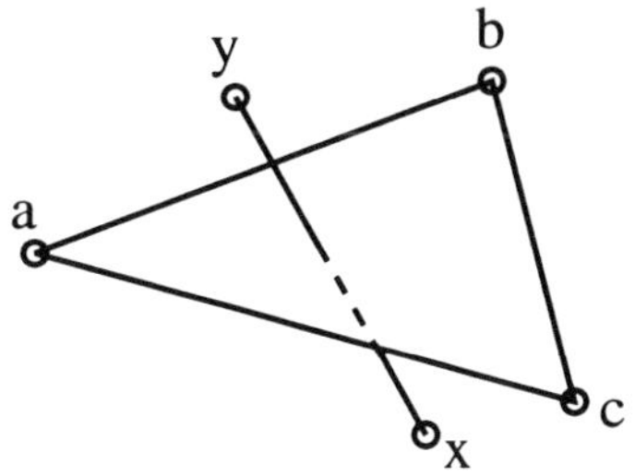

Fig. 5. Triangle and line segment. Line segment $\{x, y\}$ piercing a triangle $\{a, b, c\}$ can be expressed by signs of determinants.

We consider a triangle $\{a, b, c\}$ and a line segment $\{x, y\}$ in real 3-dimensional affine space. For simplicity, we assume a generic position of our 5 points. In order to understand an essential property, we first use coordinates. We order the 5 vertices $\{a, b, c, x, y\}$, and we write for these points homogeneous coordinates, in other words, we add a fourth coordinate for each point being 1 for all our points. We calculate all 4×4 sub-determinants formed of the 5×4 matrix of homogeneous coordinates of these points. The determinant will be denoted by $[a, b, \underline{c}, x, y]$ when for instance the point c does not occur. We see that the property of having a point of intersection is given precisely by a sign condition of the determinants. When denoting $T := sign[\underline{a}, b, c, x, y] = -sign[a, \underline{b}, c, x, y] = sign[a, b, \underline{c}, x, y]$, and $S := -sign([a, b, c, \underline{x}, y] = sign[a, b, c, x, \underline{y}]$, this condition reads $T = -S$. In problems of this kind, it is natural to ask of whether the calculations with coordinates can be avoided.

The essential condition deals with signs only.

1.5. Zonotopal tilings

We mention an aspect of oriented matroids which seems to have nothing to do with point sets at all. It can stand for the surprising variety of how oriented matroids show up. Dress observed that determining all zonotopal tilings in the plane is equivalent with the problem of determining all rank 3 oriented matroids. Those not having a matrix representative have to be included, see [6].

In the example of a regular 12-gon in Fig. 14 one can check easily the correspondence between the topological cell decomposition and its zonotopal tiling.

The very natural and still very challenging direction of research in oriented matroid theory lies in casting more light along the borderline between oriented matroids of realizable point configurations and oriented matroids in general. Harborth's problem in [28] was of this type. Fig. 14 shows an example where we see (one half of) the topological cell decomposition of the 2-sphere. It has a maximal number of triangles. Can we obtain a homeomorphic picture in which these curves are projections of great circles of the 2-sphere? A non-affirmative answer has been given in [19]. Our topological information about the cell decomposition of the 2-sphere can also be stored as a signed vector with $\binom{28}{3}$ components and vice versa: signed vectors fulfilling the oriented matroid axioms can always be drawn in this way.

1.6. Oriented matroids in the chirotope notation

For an $n \times d$-matrix M, we study the map $\operatorname{sign}(x_1 \wedge x_2 \wedge \ldots \wedge x_d)$ of the exterior product of its column vectors $x_1, \ldots, x_d$; in other words, we consider signed vectors of the form $(\ldots, \operatorname{sign}[i_1, \ldots, i_d], \ldots) \in \{-1, 0, +1\}^{\binom{n}{d}}$ where $[i_1, \ldots, i_d]$ denotes the $(d \times d)$-subdeterminant of M formed of the rows $i_1, \ldots, i_d$ of M in the given order. Any attempt of characterizing combinatorially precisely those signed vectors which arise from real matrices by providing a finite set of axioms for such signed vectors must be a failure. The missing axiom is lost forever, compare [21]. What we have instead is a natural generalization, a closed set for such signed vectors: oriented matroids.

2. The set of realizable oriented matroids and its closure

This section was chosen to illustrate the surprisingly different structures which have led to oriented matroids. In fact, it took some time during the last 20 years to understand and to proof the isomorphisms between these various forms of representing oriented matroids. Once we have gained a certain flexibility to switch between these concepts, we benefit from what has been established within another model and what otherwise cannot be seen so easily in the given model.

Natural generalizations of the concept of an oriented matroid derived from a matrix led independently to oriented matroids in the general setting. We list five properties $\mathcal{P}_1, \mathcal{P}_2, \ldots, \mathcal{P}_5$ such that looking for the closed set of mathematical objects with property $\mathcal{P}_i$ leads to the concept of oriented matroids in its general form.

2.1. Farkas' lemma from operations research

$(\mathcal{P}_1)$: **Let A be an $m \times n$ matrix and $b \in R^m$. Either there is a point $x \in R^n$ with $Ax \leq b$ or there is a point $u \in R^m_+$ with $A^t u = 0$ and $u^t b < 0$.**

Oriented matroids can be considered generalizing this duality of orthogonal linear subspaces. This was Bland's approach in his Ph.D. thesis towards oriented matroids. See Bachem and Kern [3] for a detailed exposition of oriented matroids when viewed as a tool in operations research.

2.2. Circuit properties of oriented graphs

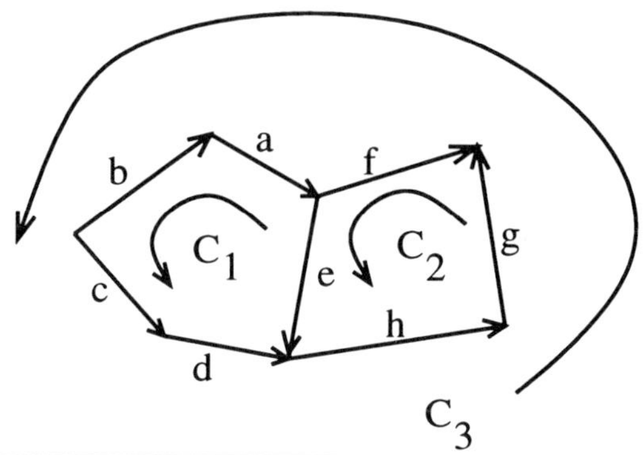

a	b	c	d	e	f	g	h	
−	−	+	+	−	0	0	0	C_1
0	0	0	0	+	−	+	+	C_2
−	−	+	+	0	−	+	+	C_3

Fig. 6. Example of oriented circuits.

($\mathcal{P}_2$) : **Given any two circuits C_1 and C_2 in an oriented graph with a common edge e, there exists a third circuit C_3 with edges from the union of C_1 and C_2 but without the common edge e.**

When looking at the circuit axioms of Las Vergnas presented first in 1974, [30], it is useful to think of this concept. Oriented matroids generalize combinatorial properties of directed graphs and convex polytopes, see Las Vergnas' papers cited in [5] or [9].

2.3. Incidence properties of lines in the real projective plane

($\mathcal{P}_3$) : **Two different lines in the projective plane cross each other exactly once where they meet.**

n closed curves in the projective plane with the same incidence properties as n lines yield a model for a *reorientation class of an oriented matroid in rank 3.* The collection of n such closed curves is called a *pseudoline arrangement.* When we pick a reference line and we assume the remaining lines to be oriented as in the next example, we have an oriented matroid of rank 3. More precisely, the oriented matroid is an equivalence class of all pseudoline arrangements having the same topological properties. On the other hand, for each oriented matroid of rank 3, it is possible to think of it as a pseudoline arrangement. Grünbaum in [27] provides a comprehensive description of pseudoline arrangements. He summarized their properties before oriented matroids of arbitrary rank were introduced. Levi in [33] was apparently the first to introduce this notion in 1926. The following example of Roudneff in Fig. 7 was studied in [34].

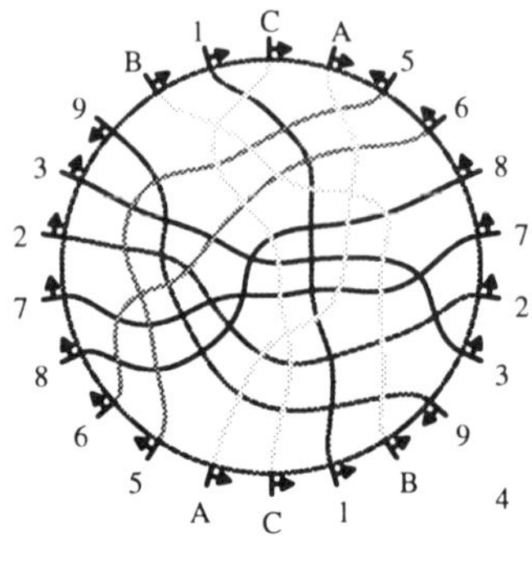

Fig. 7. Pseudoline arrangement without adjacent triangles. This example of a pseudoline arrangement has no adjacent triangles. Line 4 has been chosen as reference line.

2.4. Incidences of hyperplanes in real projective d-space

($\mathcal{P}_4$) Any two different *great* $(d-2)$-spheres on a $(d-1)$-unit sphere in R^d intersect in a $(d-3)$-sphere.

In replacing the spheres with topological spheres while requiring the intersection properties as in the metrical case, we arrive at the Edmonds-Mandel-Folkman-Lawrence axiomatic for oriented matroids.

In other words, we have generalized the concept of Subsection 2.3 to higher dimensions. For $d = 4$ we have the unit 3-sphere, the closed upper half of which we project down onto 3-space. Within this 3-ball we have a finite set of topological discs two of which intersect in a topological line where they cross each other. We can draw the topological discs. The way they intersect is known up to a mirror image because of the common topological lines for each pair of topological planes.

In general, a set of n oriented hyperplanes $H_1, \ldots, H_n$ passing through the origin o of euclidean d-space R^d, defines a cell decomposition of the $(d-1)$-sphere S^{d-1} with center at 0 as well as a decomposition of the projective $(d-1)$-space RP^{d-1} where antipodal points of S^{d-1} have to be identified. Such a decomposition is called a *sphere system* or more precisely, a representable sphere system in this case. In the uniform case any d of the oriented $(d-2)$-spheres $S_i := H_i \cap S^{d-1}$ have empty intersection.

When going along 1-spheres defined as intersections of any $d-2$ pseudoplanes of the S_i, we meet in a circular order all other $(d-2)$-spheres $S_i := H_i \cap S^{d-1}$. We look at the equivalence class of all those sphere systems having the same circular order for all these 1-spheres. This topological object is isomorphic to a *realizable oriented matroid* . Moreover, we get the general case of an arbitrary oriented matroid as an equivalence class of sphere systems according to Edmonds, Mandel, Folkman, and Lawrence, whereby we replace the oriented $(d-2)$-spheres with topological spheres, and we require intersection properties similar to the representable case.

2.5. On the Bielefeld pseudoplane arrangement

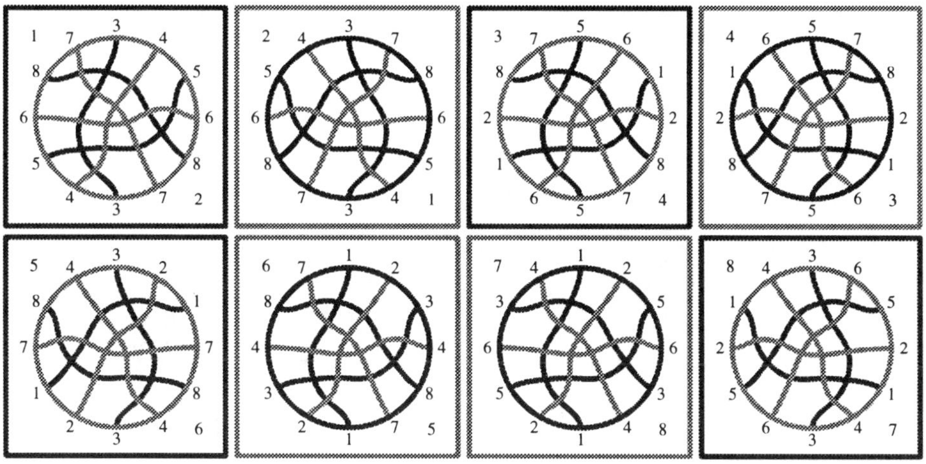

Fig. 8. Reorientation class in rank 4 with 8 elements and 7 mutations.

The chosen example in Fig. 8 shows a smallest non-representable pseudo-plane arrangement in projective 3-space with 8 pseudo-planes which has the smallest number 7 of tetrahedral cells (this making it unique), see Bokowski and Richter-Gebert [16]. There exists a model which was produced by the author for the Center of Interdisciplinary Reseach (ZIF), Bielefeld, see a coloured picture in [9]. The example was known to Roudneff and, independently, it formed the essential contradiction why a particular sphere of Altshuler (investigated by the author) cannot be realized as the boundary of a convex polytope. Starting with a tetrahedron, this oriented matroid can be described as follows. Eight (oriented) planes are defined by the affine hull of the 4 facets of the given tetrahedron together with the affine hull of the 4 facets of its homothetic outer tetrahedron. Towards the plane at infinity, these 8 (oriented) planes are deformed to pseudo-planes in order to have a *simple arrangement of pseudo-planes,* ie only 3 pseudo-planes meet in a common point. When the tetrahedral symmetry is maintained, we see for each edge of the simplex a tetrahedral region having a 4-gon as intersection with the plane at infinity. Together with the central tetrahedron, we have 7 tetrahedral cells altogether.

2.5.1. *The regular 24-cell*

We analyse now the regular 24-cell as an example. The homogeneous coordinates can be written as a 24x5-matrix $\underline{M}$. When we substract the first column from the second, the oriented matroid of the matrix does not change (rules for determinants). The projection from the first point onto a 3-dimensional plane means (because of the zeros) deleting the first column and the first row in order to obtain the next matrix $\underline{N}$ with 23 elements. We label the rows in the ordering as indicated The so called contraction at the next point (at the first row vector of $\underline{N}$) can be obtained again (because of the zeros) by deleting the first row and the first column. We see that there are only 7 essentially different normal vectors in the matrix $\underline{L}$.

$$\underline{M} = \begin{pmatrix} 2 & 2 & 0 & 0 & 0 \\ 2 & -2 & 0 & 0 & 0 \\ 2 & 0 & 2 & 0 & 0 \\ 2 & 0 & -2 & 0 & 0 \\ 2 & 0 & 0 & 2 & 0 \\ 2 & 0 & 0 & -2 & 0 \\ 2 & 0 & 0 & 0 & 2 \\ 2 & 0 & 0 & 0 & -2 \\ 2 & 1 & 1 & 1 & 1 \\ 2 & -1 & 1 & 1 & 1 \\ 2 & 1 & -1 & 1 & 1 \\ 2 & -1 & -1 & 1 & 1 \\ 2 & 1 & 1 & -1 & 1 \\ 2 & -1 & 1 & -1 & 1 \\ 2 & 1 & -1 & -1 & 1 \\ 2 & -1 & -1 & -1 & 1 \\ 2 & 1 & 1 & 1 & -1 \\ 2 & -1 & 1 & 1 & -1 \\ 2 & 1 & -1 & 1 & -1 \\ 2 & -1 & -1 & 1 & -1 \\ 2 & 1 & 1 & -1 & -1 \\ 2 & -1 & 1 & -1 & -1 \\ 2 & 1 & -1 & -1 & -1 \\ 2 & -1 & -1 & -1 & -1 \end{pmatrix} \qquad \underline{N} = \begin{array}{c} M \\ H \\ J \\ I \\ K \\ L \\ N \\ 1 \\ 9 \\ 2 \\ A \\ 3 \\ B \\ 4 \\ C \\ 5 \\ D \\ 6 \\ E \\ 7 \\ F \\ 8 \\ G \end{array} \begin{pmatrix} -4 & 0 & 0 & 0 \\ -2 & 2 & 0 & 0 \\ -2 & -2 & 0 & 0 \\ -2 & 0 & 2 & 0 \\ -2 & 0 & -2 & 0 \\ -2 & 0 & 0 & 2 \\ -2 & 0 & 0 & -2 \\ -1 & 1 & 1 & 1 \\ -3 & 1 & 1 & 1 \\ -1 & -1 & 1 & 1 \\ -3 & -1 & 1 & 1 \\ -1 & 1 & -1 & 1 \\ -3 & 1 & -1 & 1 \\ -1 & -1 & -1 & 1 \\ -3 & -1 & -1 & 1 \\ -1 & 1 & 1 & -1 \\ -3 & 1 & 1 & -1 \\ -1 & -1 & 1 & -1 \\ -3 & -1 & 1 & -1 \\ -1 & 1 & -1 & -1 \\ -3 & 1 & -1 & -1 \\ -1 & -1 & -1 & -1 \\ -3 & -1 & -1 & -1 \end{pmatrix} \qquad \underline{L} = \begin{array}{c} H \\ \\ I \\ \\ L \\ \\ 1 \\ \\ 2 \\ \\ 3 \\ \\ 4 \\ \\ \\ \\ \\ \\ \\ \\ \\ \\ \\ \\ \end{array} \begin{pmatrix} 2 & 0 & 0 \\ -2 & 0 & 0 \\ 0 & 2 & 0 \\ 0 & -2 & 0 \\ 0 & 0 & 2 \\ 0 & 0 & -2 \\ 1 & 1 & 1 \\ 1 & 1 & 1 \\ -1 & 1 & 1 \\ -1 & 1 & 1 \\ 1 & -1 & 1 \\ 1 & -1 & 1 \\ -1 & -1 & 1 \\ -1 & -1 & 1 \\ 1 & 1 & -1 \\ 1 & 1 & -1 \\ -1 & 1 & -1 \\ -1 & 1 & -1 \\ 1 & -1 & -1 \\ 1 & -1 & -1 \\ -1 & -1 & -1 \\ -1 & -1 & -1 \end{pmatrix}$$

The remaining row vectors of $\underline{L}$ are multiples of these row vectors. This matrix $\underline{L}$ corresponds to one example (in Fig. 15 it is the middle one in the bottom row) of the complete set of 23 rank 3 contractions which altogether define the topological invariant of the vertex figure of the 24-cell: its contraction at one vertex. The corresponding topological invariant is again a three dimensional one with all rank 3 contractions glued along common pseudolines where they cross, compare the colored Fig. 15. Symmetry aspects can be seen immediately from studying these invariants. These pictures have been used in [14] in order to show that the reorientation class of the matroid of the 24-cell is unique.

2.6. DETERMINANT SIGN CONDITIONS

$(\mathcal{P}_5)$: Signs of $d \times d$-subdeterminants of $n \times d$-matrices do not violate Grassmann-Plücker relations.

For the smallest interesting example, we consider the two-dimensional vectorspace $V \subset W = R^4$ with basis (x,y), $x = (x_1, x_2, x_3, x_4)^t = (1, 0, a, c)^t$, $y = (y_1, y_2, y_3, y_4)^t = (0, 1, b, d)^t$. With $[i,j] := \det \begin{pmatrix} x_i & y_i \\ x_j & y_j \end{pmatrix}$, the oriented matroid $\chi = \chi(V) \in \{-1, 0, 1\}^{\binom{dim W}{dim V}}$ of 4 points in rank 2 is defined as

$$\chi(V) = sign(x \wedge y) =$$

$$= (sign[1,2],\ sign[1,3],\ sign[1,4],\ sign[2,3],\ sign[2,4],\ sign[3,4]).$$

The relation for determinants (Grassmann-Plücker relation)

$$\det \begin{pmatrix} 1 & 0 \\ 0 & 1 \end{pmatrix} \cdot \det \begin{pmatrix} a & b \\ c & d \end{pmatrix} - \det \begin{pmatrix} 1 & 0 \\ a & b \end{pmatrix} \cdot \det \begin{pmatrix} 0 & 1 \\ c & d \end{pmatrix} + \det \begin{pmatrix} 1 & 0 \\ c & d \end{pmatrix} \cdot \det \begin{pmatrix} 0 & 1 \\ a & b \end{pmatrix} = 0,$$

$$\text{or} \qquad (**) \qquad \det[1,2]\det[3,4] - \det[1,3]\det[2,4] + \det[1,4]\det[2,3] = 0,$$

leads to a sign condition $(*)$ the signs of questionable determinants have to fulfil:

$$(*) \qquad A + B + C + ABC = 0, \qquad A, B, C \in \{-1, 0, 1\},$$

$$\begin{aligned} A &:= sign[1,2]sign[3,4], \\ B &:= -sign[1,3]sign[2,4], \\ C &:= sign[1,4]sign[2,3]. \end{aligned}$$

The determinant signs

$$(\text{sign}[1,2], \text{sign}[3,4], \text{sign}[1,3], \text{sign}[2,4], \text{sign}[1,4], \text{sign}[2,3])$$

do not violate the Grassmann-Plücker relation $(**)$ iff the sign condition $(*)$ holds, compare [13]. A *signed vector* $\mathcal{X} \in \{-1,0,1\}^{\binom{n}{d}}$ is an oriented matroid if and only if no Grassmann-Plücker relation is violated in the above sense.

We consider now the general case. The d-dimensional vectorspace $V \in R^n$. can be written in the form $V = \{\sum_{i=1}^{d} p_i c_i : p_i \in \mathbf{R}\}$ with $(c_1, c_2, \ldots, c_d)$ denoting a basis of $V \subset \mathbf{R}^n$. We think of $(c_1, c_2, \ldots, c_d)$ also as column vectors of an $n \times d$-matrix. We consider our given vectorspace V as a point p on the Grassmann-manifold. We use the exterior algebra of an n-dimensional vectorspace $\mathbf{R}^n$ and the notion of Grassmann coordinates, ie simple (or decomposable) d-vectors in the exterior product space $\wedge_d \mathbf{R}^n$, or $R^{\binom{n}{d}}$. We consider the Grassmann manifold $\mathcal{G}_{n,d}^{\mathbf{R}}$ over $\mathbf{R}$ of all d-dimensional (unoriented) subspaces of $\mathbf{R}^n$. A point p on the Grassmann manifold, representing such a subspace, and represented by a simple d-vector in $\wedge_d \mathbf{R}^n$, can be used to define an oriented matroid in the realizable case.

Definition. (Realizable oriented Matroid) An *oriented matroid* $\mathcal{X}$ with n points of rank d on $E = \{1, 2, \ldots, n\}$ is a map $\mathcal{X} : E^d \to \{-1, 0, +1\}$ with $\mathcal{X} = sign(p)$ for some point p on the Grassmann manifold.
We write $\Lambda(n,d)$ for the set of all coordinates

$$[\lambda] = [\lambda_1, \lambda_2, \ldots, \lambda_d] \in E^d, 1 \le \lambda_1 < \lambda_2 < \cdots < \lambda_d \le n,$$

(*bracket variables*) in $\mathbf{R}^{\binom{n}{d}}$. These coordinates define similar expressions $[\pi(\lambda)] = [\lambda_{\pi(1)}, \lambda_{\pi(2)}, \ldots, \lambda_{\pi(d)}]$ for all permutations π of the d numbers $\lambda_1, \lambda_2, \ldots, \lambda_d$, $[\pi(\lambda)] = \text{sign}(\pi) \cdot [\lambda]$. We define $[\lambda_1, \ldots, \lambda_d] := 0$ if for any $1 \le i \ne j \le n$, $\lambda_i = \lambda_j$.

Given an $n \times d$-matrix, its $d \times d$-subdeterminants form the coordinates of a point in the space $\mathbf{R}^{\binom{n}{d}}$. The set of all these points for all real matrices form the *Grassmannian* . This variety can be expressed via the dependencies between the $d \times d$-subdeterminants. They can be expressed in the form

$$\sum_{i=1}^{d+1} (-1)^i [\lambda_1, \ldots, \widehat{\lambda_i}, \ldots, \lambda_{d+1}][\mu_1, \ldots, \mu_{d-1}, \lambda_i] = 0.$$

The widehat denotes that this element has to be discarded. In the uniform case, an oriented matroid can be defined to be a mapping $\mathcal{X} \colon \Lambda(n,d) \to \{-1,+1\}$ with the following property for its unique alternating extension $\mathcal{X} \colon \{1, \ldots, n\}^d \to \{-1,+1\}$: for all $\sigma \in \Lambda(n, d-2)$ and for all $\tau \in \Lambda(n, 4)$, the set

$$\begin{aligned} \{\, &\mathcal{X}(\ldots, \sigma_{d-2}, \tau_1, \tau_2) \cdot \mathcal{X}(\ldots, \sigma_{d-2}, \tau_3, \tau_4), \\ -&\mathcal{X}(\ldots, \sigma_{d-2}, \tau_1, \tau_3) \cdot \mathcal{X}(\ldots, \sigma_{d-2}, \tau_2, \tau_4), \\ &\mathcal{X}(\ldots, \sigma_{d-2}, \tau_1, \tau_4) \cdot \mathcal{X}(\ldots, \sigma_{d-2}, \tau_2, \tau_3) \,\} \end{aligned}$$

either contains $\{-1, +1\}$ or equals $\{0\}$.

3. Oriented matroids of rank d defined via hyperline sequences

We fix a set of n abstract points $E := \{1, 2, \ldots, n\}$ and a number $d \geq 3$. For any ordered set $L := \{\sigma_1, \ldots, \sigma_{d-2}\}$ of $d-2$ numbers with $\sigma_1 < \ldots < \sigma_{d-2}$, $\sigma_i \in E$, we assume to have given a sequence of the form

$$(\sigma_1, \ldots, \sigma_{d-2} \mid \operatorname{sign}(\tau_1) \cdot \tau_1, \ldots, \operatorname{sign}(\tau_{n-d+2}) \cdot \tau_{n-d+2})$$

with $\tau_j \in E \setminus L$ occuring exactly once.
Think of the topological lines used in Subsection 2.3 and Subsection 2.4 for which the sequence how the pseudoplanes intersect these lines has been indicated.
For the sign of the tupel $(\sigma_1, \ldots, \sigma_{d-2}, \tau_j, \tau_k)$, we consider the sequence starting with $\sigma_1, \ldots, \sigma_{d-2}$. We define the sign of the formal variable (questionable determinant) $[\sigma_1, \ldots, \sigma_{d-2}, \tau_j, \tau_k]$ to be $+1$ when $\operatorname{sign}(\tau_j) = \operatorname{sign}(\tau_k)$ and when τ_j comes before τ_k in the above sequence. We require that both, the opposite case of the position of points and opposite signs, cause sign changes of the formal determinant $[\sigma_1, \ldots, \sigma_{d-2}, \tau_j, \tau_k]$. If and only if this definition for the signs of formal determinants is consistent and does not violate the alternating rule for determinants, we define the above set of sequences to be *a uniform oriented matroid with n points of rank d*. Oriented matroids can be stored this way for computational purposes.

3.1. Geometric interpretation of hyperline sequences

For a finite set of points in affine real $(d-1)$-space, we can consider the oriented part $L := \{\sigma_1, \ldots, \sigma_{d-2}\}$ of the sequence as an oriented hyperline which induces an orientation in the two-dimensional plane orthogonal to it. When rotating an oriented hyperplane continuously around L according to the orientation of L (or its orthogonal complement), it will meet the points in a certain sequence just as in the formal definition above. The signs indicate of whether the points lie on the positive or on the negative side of the oriented hyperplane just before they are incident with it. This concept of the author was applied in investigations of questionable polytopes already in [2]. It is closely related to the allowable sequence concept of Goodman and Pollack, [26]. The condition of not violating the alternating rule for determinants is equivalent to the sign condition (*) for determinants.

3.2. Example of signed sequences for a regular 15-gon.

Sometimes the signed sequences are not the essential structure but only the order in which the intersections occur. In Fig. 9 we have the following sequences.

(1	|	2	A	3	B	4	C	5	D	6	E	7	F	8	G	9)
(2	|	1	A	B	C	D	E	F	G	3	4	5	6	7	8	9)
(3	|	1	B	C	D	E	F	G	2	4	5	6	7	8	9	A)
(4	|	1	C	D	E	F	G	2	3	5	6	7	8	9	A	B)
(5	|	1	D	E	F	G	2	3	4	6	7	8	9	A	B	C)
(6	|	1	E	F	G	2	3	4	5	7	8	9	A	B	C	D)
(7	|	1	F	G	2	3	4	5	6	8	9	A	B	C	D	E)
(8	|	1	G	2	3	4	5	6	7	9	A	B	C	D	E	F)
(9	|	1	2	3	4	5	6	7	8	A	B	C	D	E	F	G)
(A	|	1	3	4	5	6	7	8	9	B	C	D	E	F	G	2)
(B	|	1	4	5	6	7	8	9	A	C	D	E	F	G	2	3)
(C	|	1	5	6	7	8	9	A	B	D	E	F	G	2	3	4)
(D	|	1	6	7	8	9	A	B	C	E	F	G	2	3	4	5)
(E	|	1	7	8	9	A	B	C	D	F	G	2	3	4	5	6)
(F	|	1	8	9	A	B	C	D	E	G	2	3	4	5	6	7)
(G	|	1	9	A	B	C	D	E	F	2	3	4	5	6	7	8)

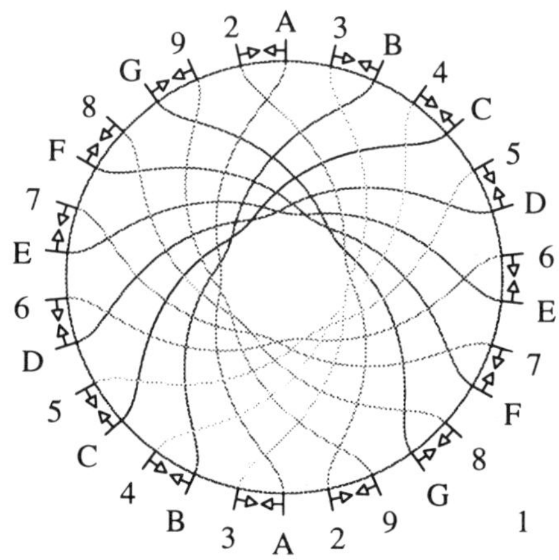

Fig. 9. The oriented matroid corresponding to a regular 15-gon with line at infinity.

3.3. Pseudoplane arrangement with 10 elements and 9 simplicial cells.

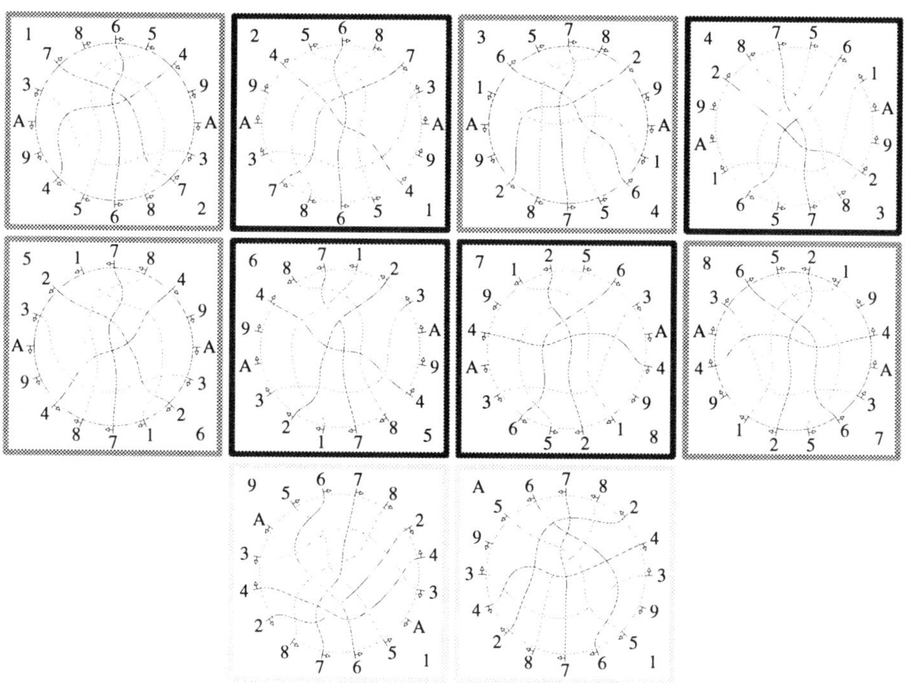

Fig. 10. Oriented matroid in rank 4 with 10 elements and 9 mutations.

We consider an oriented matroid with 10 elements in rank 4. We provide the sphere system model as before by listing all 10 *contractions*, ie the 10 pseudoplanes together with the information how the others intersect. The combinatorial description as a set of hyperline sequences is very useful for algorithmical purposes. It can be derived from the pictures in a straightforward way. One only has to proceed along the pseudolines of intersections and to write down the pseudolines together with its orientations when they are crossed.

Hyperline Sequences of the arrangement:

$(1,2\vert +3,+A,+9,+4,+5,+6,+8,+7)$	$(3,5\vert -1,-2,-7,-8,+4,+9,+A,-6)$
$(1,3\vert -2,+4,+A,+9,+5,+6,+7,+8)$	$(3,6\vert -1,-2,-7,-8,+4,+A,+9,+5)$
$(1,4\vert -2,-9,-A,-3,+5,+6,+7,+8)$	$(3,7\vert -1,-2,+6,+5,-8,+4,+9,+A)$
$(1,5\vert -2,-3,-A,-9,-4,+7,+8,+6)$	$(3,8\vert -1,-2,+6,+5,+7,+4,+9,+A)$
$(1,6\vert -2,-3,-A,-9,-4,+7,+8,-5)$	$(4,5\vert -1,-9,-2,-7,-8,-A,-3,-6)$
$(1,7\vert -2,-8,-3,-A,-9,-4,-6,-5)$	$(4,6\vert -1,-9,-2,-7,-8,-A,-3,+5)$
$(1,8\vert -2,+7,-3,-A,-9,-4,-6,-5)$	$(4,7\vert -1,-2,-9,+6,+5,-8,-A,-3)$
$(2,3\vert -1,-A,-9,-4,-5,-6,-7,-8)$	$(4,8\vert -1,-2,-9,+6,+5,+7,-A,-3)$
$(2,4\vert -1,+3,+A,-5,-6,+9,-7,-8)$	$(5,6\vert -1,-2,-3,-A,-9,-4,+8,+7)$
$(2,5\vert -1,+3,+A,+4,+9,-7,-8,-6)$	$(5,7\vert -1,-2,-9,-4,-A,-3,-6,-8)$
$(2,6\vert -1,+3,+A,+4,+9,-7,-8,+5)$	$(5,8\vert -1,-2,-9,-4,-A,-3,-6,+7)$
$(2,7\vert -1,+8,+3,+A,+9,+4,+6,+5)$	$(6,7\vert -1,-2,-9,-4,-A,-3,+5,-8)$
$(2,8\vert -1,-7,+3,+A,+9,+4,+6,+5)$	$(6,8\vert -1,-2,-9,-4,-A,-3,+5,+7)$
$(3,4\vert -1,-A,-9,-2,-8,-7,-5,-6)$	$(7,8\vert -1,-9,-4,-A,-3,-6,-5,+2)$

We have listed only those hyperline sequences which allow us to read off all signs of bases. Hyperlines with elements 9 and A have been discarded. This example (with deleted elements 9 and A) with 8 elements was described under Fig. 8. It is also recommended to calculate the signs of formal determinants from the hyperline configuration and vice versa. Here are the signs of bases of the oriented matroid with 8 elements of Fig. 8.

Signs of Bases:

sign[1,2,3,4] = +1	sign[1,2,3,5] = +1	sign[1,2,3,6] = +1	sign[1,2,3,7] = +1
sign[1,2,3,8] = +1	sign[1,2,4,5] = +1	sign[1,2,4,6] = +1	sign[1,2,4,7] = +1
sign[1,2,4,8] = +1	sign[1,2,5,6] = +1	sign[1,2,5,7] = +1	sign[1,2,5,8] = +1
sign[1,2,6,7] = +1	sign[1,2,6,8] = +1	sign[1,2,7,8] = −1	sign[1,3,4,5] = +1
sign[1,3,4,6] = +1	sign[1,3,4,7] = +1	sign[1,3,4,8] = +1	sign[1,3,5,6] = +1
sign[1,3,5,7] = +1	sign[1,3,5,8] = +1	sign[1,3,6,7] = +1	sign[1,3,6,8] = +1
sign[1,3,7,8] = +1	sign[1,4,5,6] = +1	sign[1,4,5,7] = +1	sign[1,4,5,8] = +1
sign[1,4,6,7] = +1	sign[1,4,6,8] = +1	sign[1,4,7,8] = +1	sign[1,5,6,7] = −1
sign[1,5,6,8] = −1	sign[1,5,7,8] = +1	sign[1,6,7,8] = +1	sign[2,3,4,5] = +1
sign[2,3,4,6] = +1	sign[2,3,4,7] = +1	sign[2,3,4,8] = +1	sign[2,3,5,6] = +1
sign[2,3,5,7] = +1	sign[2,3,5,8] = +1	sign[2,3,6,7] = +1	sign[2,3,6,8] = +1
sign[2,3,7,8] = +1	sign[2,4,5,6] = +1	sign[2,4,5,7] = +1	sign[2,4,5,8] = +1
sign[2,4,6,7] = +1	sign[2,4,6,8] = +1	sign[2,4,7,8] = +1	sign[2,5,6,7] = −1
sign[2,5,6,8] = −1	sign[2,5,7,8] = +1	sign[2,6,7,8] = +1	sign[3,4,5,6] = +1
sign[3,4,5,7] = −1	sign[3,4,5,8] = −1	sign[3,4,6,7] = −1	sign[3,4,6,8] = −1
sign[3,4,7,8] = −1	sign[3,5,6,7] = −1	sign[3,5,6,8] = −1	sign[3,5,7,8] = +1
sign[3,6,7,8] = +1	sign[4,5,6,7] = −1	sign[4,5,6,8] = −1	sign[4,5,7,8] = +1
sign[4,6,7,8] = +1	sign[5,6,7,8] = −1		

4. On the Las Vergnas conjecture concerning simplicial cells

We denote the number of $(d-1)$-dimensional simplicial cells in the projective $(d-1)$-space of an oriented matroid with n elements in rank d by $sc(n,d)$. In the representable case, Shannon [40] showed $sc(n,d) \geq n$. This result is also best possible in the representable case with the alternating oriented matroid (cyclic polytope) as extremum.

Conjecture (Las Vergnas) Every uniform pseudoplane arrangement ($n \geq d$) has at least one simplicial cell: $sc(n, d) \geq 1$, $d \geq 4$.

The case $d = 3$ is rather trivial and in fact, we find the answer already in Levi's paper from 1926, [33]. Here we study non-trivial cases in rank 4, $d = 4$, $sc(n) := sc(n, 4)$.

The classification in [16] was motivated in part by looking for examples with $sc(n) < n$. But it has turned out that there is up to reorientation just a single case with $sc(n) < n$ for $n = 8$.

The surprising fact that there was just a single example known with $sc(n) < n$ for $n = 8$ led to the question of whether Shannon's result can be used for other examples, thus showing non-representability by just calculating the number of simplicial cells. The following theorem answers this in the affirmative.

Theorem. For $n > 7$ there exists at least one reorientation class with $sc(n) = n - 1$.

The unique example with 7 simplicial cells can be extended for instance to get the example with 10 elements of Fig. 10. All 10 contractions of this example are shown. Deleting A leads to an example with 8 simplicial cells and 9 elements, and deleting 9 leads to the unique example with 7 simplicial cells and 8 elements.

Here we can give an affirmative answer to Las Vergnas' conjecture for all oriented matroids in rank 4 with up to 12 elements.

Theorem A counter-example to Las Vergnas' conjecture must have at least 13 elements, cf. [11].

Remark 1. The classification of all 312.356 reorientation classes of 10 point oriented matroids in rank 3 into representable and non-representable ones in [15] has shown that an overview about the non-representable cases with 13 elements in rank 4 seems to be without reach.

Remark 2. A duality argument yields results also for cases of higher rank.

5. Collection of small examples

The headline of this section can be viewed as the headline for the whole article. We have already presented several examples which played a role in the theory of oriented matroids. We hope that the reader gets a better understanding of the theory of oriented matroids via working with examples. The program OMAWIN (on X-WINDOWS) reduces the boring work of handling these examples a lot. When the reader wants to work with oriented matroids, he will find it very convenient to have interesting small examples for various purposes, especially non-realizable ones. For instance one can test and sometimes find a contradiction to a property well known for point configurations which fails to hold on the level of oriented matroids.

5.1. Matroids versus oriented matroids

When we compare the matroid structure with the structure of oriented matroids, the transition from the sign map for all bases of an oriented matroid to the map of its absolute values changes an oriented matroid $\mathcal{M}$ to its underlying matroid $\underline{\mathcal{M}}$. This is the easy direction. To see this, one can look at the Grassmann-Plücker relations in such a way that the Steinitz-Mac Lane exchange lemma is seen. For a pair of bases of the underlying matroid one can form a Grassmann-Plücker relation with the corresponding brackets of these bases as a summand in the polynomial. The oriented matroid requires that there must exist another non-vanishing summand. This amounts to the Steinitz-Mac Lane exchange lemma.

5.1.1. *Example of an oriented matroid in terms of its signed bases*

$$\begin{array}{llll}
\mathrm{sign}[1,2,3] = 0 & \mathrm{sign}[1,2,4] = -1 & \mathrm{sign}[1,2,5] = -1 & \mathrm{sign}[1,2,6] = -1 \\
\mathrm{sign}[1,3,4] = -1 & \mathrm{sign}[1,3,5] = -1 & \mathrm{sign}[1,3,6] = -1 & \mathrm{sign}[1,4,5] = -1 \\
\mathrm{sign}[1,4,6] = -1 & \mathrm{sign}[1,5,6] = 0 & \mathrm{sign}[2,3,4] = -1 & \mathrm{sign}[2,3,5] = -1 \\
\mathrm{sign}[2,3,6] = -1 & \mathrm{sign}[2,4,5] = -1 & \mathrm{sign}[2,4,6] = 0 & \mathrm{sign}[2,5,6] = +1 \\
\mathrm{sign}[3,4,5] = 0 & \mathrm{sign}[3,4,6] = +1 & \mathrm{sign}[3,5,6] = +1 & \mathrm{sign}[4,5,6] = +1
\end{array}$$

The input for the program OMAWIN in this case reads

6	= number of elements
3	= rank
0 − − − − − − − − −0 − − − − −0 + 0 + ++	= signed bases ordered as above.

Its output can be seen in Fig. 11, orientations have not been indicated.

Fig. 11. Example with different choices for the line at infinity.

5.1.2. *Example of a matroid in terms of its bases*

	[1, 2, 4]	[1, 2, 5]	[1, 2, 6]
[1, 3, 4]	[1, 3, 5]	[1, 3, 6]	[1, 4, 5]
[1, 4, 6]		[2, 3, 4]	[2, 3, 5]
[2, 3, 6]	[2, 4, 5]		[2, 5, 6]
	[3, 4, 6]	[3, 5, 6]	[4, 5, 6]

One way of comparing matroids with oriented matroids can be seen in [13], see also contributions in [10], [14].

5.2. All non-realizable reorientation classes of oriented matroids, 8 elements, rank 4.

The following table contains representatives of all 24 non-realizable reorientation classes of uniform oriented matroids in rank 4 with 8 elements, [16]. The OMAWIN program can transform each of them into a picture like Fig. 8.

```
1   +++++----------++++++-+----+-++++++------+-++++-+-------++++-++++----+
2   --++++---+++-+++-----+---++++++---+-+++--++--++-+++---+++-+++---+----+-
3   ++---++++++--+-++++++---+--++++++-+------++--++--+---+++-+++++----+-
4   +-++-+--+-++-+-+--+---++++++-----+-++--+--+++--+-+--+-+-++--++-+---+--
5   -++--++---+++++----+++++-++++-++++--+++-------++++++++--++++++++-++---
6   +++++---------+++++++++++------++-------------+++++--++-++++++++----++-
7   +++++---------+++++++++----+-++++++---+--+-++++-+-------+++++++++-+---
8   +--++----++++++--+-+++++-+-+--+++--+-+++--+-++-----++++-++-+++-+-+--+
9   ++-+++-------+++---+-+++++--+-------++-++--+-++++++-+++++-+--++++-----
10  +--++----+++++++--+--++++++++-++++++++-----++-+--+++-++++-+++++++++
11  --+------+--++-+---+--++++--+-------+-+------++-+---+----++---------+---++
12  --++-++--+--+-++-+--+--+++--++-++-++++++++--------+--+--+-++-++-+--+++
13  -+--+-+-+-+--++--++++-+++-+-+-+++++++--+-+++++-+++--+++++-----++--+
14  -+--+++++---++++-+---++-++++-++---+--++++-++++-----+++++--++-+-+---++-
15  ++++---+++--+++++-++-+-+--+---+--+++----+--+-+-+++--------+----++++
16  ++++---+--++-+---+--++++--++-----+---+--++---+++-++++--+++++++++-+++-+
17  ----++++---+-+----+-+-+-++---+-+-++---+--++-+++-+--+-+++-+-++-+--+-+++
18  ----+-+++++++++-+--++----+--++++++---+-++-+-+----++-+++-++++--+--++-++
19  ++++---+--+-+----+--+-+-++-+++-+-++++-++-+----++-++-+++++++-+++---++
20  ++++-------------+--+-+-++-+------+++-++++++-+++++++-++-+-+---------
21  ++++-------------+--+---++-+------+++-++-+++-+++++++-++-+-+-----+---
22  ++++-------------+--+-+-++-+------+++-++++++-+++++++-++-+-+-----+---
23  +++++---+--+-++----------++++++-++++++++++++-----+--+--++++++--+++--+++
24  +++++--++-+++++-----++++-++++-++--++-+----++----+++++-++++-++-+++-++
```

When handling any non-trivial examples, there are some boring calculations involved when switching from one model of representing the oriented matroid to the other. In order to reduce these calculations as much as possible, various attempts have been made to write software about oriented matroids. These programs have been very often research oriented, and their purpose was just to support a particular problem. The Darmstadt contributions tried to be more independent of the particular problem. See [20] for a first attempt in this direction. New research oriented approaches implemented by many students are available from Darmstadt. (eg the OMAWIN-program) which took into account many former experiences. Most pictures in this article are computer drawings which have been generated automatically by just having the list of oriented bases or a matrix as input. RADONCHI and CHIMAN written by P. Schuchert are other very powerful tools for generating oriented matroids with special properties and for handling oriented matroids.

5.3. The Goodman–Pollack bad pentagon

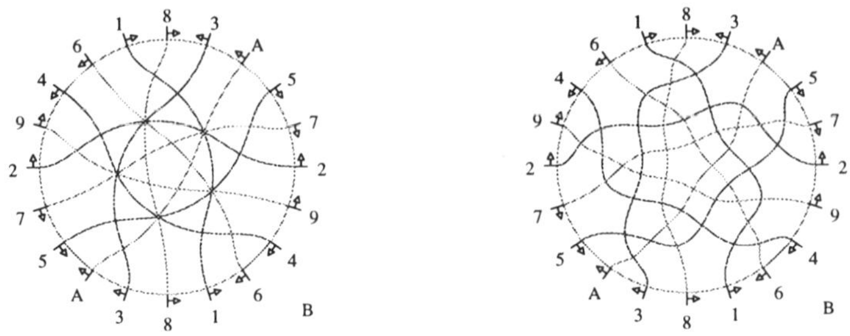

Fig. 12. Bad pentagon of Goodman and Pollack and a uniform version of it.

The example presented in Fig. 12 has been studied by Goodman and Pollack in connection with allowable sequences and later in [21] for proving that the missing axiom yielding precisely realizable oriented matroids is lost forever where it forms the first case of an infinite sequence of related examples. It can be obtained from the vertex figure of the icosahedron by changing in a cyclic way 5 signs of bases. Compare the right part in Fig. 2 with Fig. 12.

6. Abstract polyhedra

This section can be viewed as part of computational synthetical geometry, see [22] and [10]. The objective of computational synthetic geometry is finding geometrical objects or proving their non-existence when only partial, especially combinatorial data of the object is given. Here our objectives do not lie in these examples themselves but in the methods involved.

6.1. Matroid polyhedra

What is a matroid polytope? When you start with a convex polytope P, take its polar dual P^*, and its set of supporting hyperplanes defines the Folkman-Lawrence representation of the oriented matroid of the vertex set of the polytope P. This oriented matroid is called a *matroid polytope*. We get a matroid polytope in general when we have an oriented matroid such that there is a cell of maximal dimension within the Folkman-Lawrence representation such that all elements define some maximal dimensional part of the boundary of this cell. This topological cell has the boundary structure of the polar dual of a questionable polytope.

When generalizing the concept of a convex polytope to the oriented matroid setting, the notion of convexity carries over to oriented matroids as an inner property, [31]. The more general property of (not neccessarily convex) polyhedra cannot be captured by the oriented matroid itself. In this case we add the cell complex structure to the oriented matroid, and we require both structures to be compatible. We are going to make this more precise by defining an abstract notion of polyhedra. For the sake of simplicity, we deal with the 3-dimensional uniform case. We assume the polyhedra to have only triangular faces.

A *matroid polyhedron* with n points $\{1, 2, \ldots, n\}$ in rank 4 is a pair $(\mathcal{X}, \mathcal{C})$ of an oriented matroid $\mathcal{X}$ together with a simplicial cell complex $\mathcal{C}$ where the labeled abstract points and the abstract dimensions coincide for $\mathcal{X}$ and $\mathcal{C}$, and the following abstract self-intersection property is forbidden. For any triangle $\{a, b, c\}$ and any line segment $\{x, y\}$ in the cell complex $\mathcal{C}$, we do not have the following sign pattern within the set of signed bases of the oriented matroid $\mathcal{X}$.

$$\mathcal{X}[b, c, x, y] = -\mathcal{X}[a, c, x, y] = \mathcal{X}[a, b, x, y] = \mathcal{X}[a, b, c, y] = -\mathcal{X}[a, b, c, x]$$

Compare the condition $T = -S$ in the first section!

A matroid polytope $\mathcal{X}$ is an oriented matroid where all vertices of $\mathcal{X}$ are extreme. We assume again the simplicial case. A matroid polytope $\mathcal{X}$ can be viewed as a matroid polyhedron $(\mathcal{X}, \mathcal{C})$ as well. In this case the simplicial cell complex $\mathcal{C}$ is defined via the oriented matroid $\mathcal{C} := \mathcal{C}(\mathcal{X})$ itself. It makes sense to define symmetries of matroid polyhedra when the permutation group acting on the vertices respects both, the oriented matroid and the cell complex.

For the concept of *Radon partitions*, it is useful to start with a point set with homogeneous coordinates. In the above case, one has to think of points in 3-space but embedded in real 4-space. 5 vectors in R^4 are linear dependent. In other words, the zero vector can be expressed in terms of a non-trivial linear combination. In our uniform case, all points have non-zero coefficients. We partition these 5 points into two sets, according to the sign of the coefficient in the above linear combination. This is refered to as a *Radon partition* of the 5 points. In the 4-dimensional case there are essentially 3 ways to partition a set of 5 points in this way. When starting with an affine point set only two cases arise.

It turns out that this notion of Radon partitions carries over to the oriented matroid setting and indeed can be used to characterize oriented matroids.

The information about all these Radon partitions (circuit information) (minimal dependent sets) is another way to store the information of an oriented matroid. Circuits and cocircuits can be defined in an abstract manner, see [5].

6.2. A question of Sarkaria

In the following example of a matroid polyhedron $(\mathcal{X}, \mathcal{C})$, we start with the simplicial complex $\mathcal{C}$, a combinatorial torus, given by the following list of 16 triangles.

$$
\begin{array}{cccc}
(1,3,7) & (3,4,7) & (3,4,5) & (2,4,5) \\
(1,2,5) & (1,2,7) & (2,6,7) & (4,6,7) \\
(1,4,6) & (1,4,8) & (2,4,8) & (1,3,8) \\
(2,3,8) & (2,3,6) & (3,5,6) & (1,5,6)
\end{array}
$$

Sarkaria was interested how one can find a realization. It is the method we stress here. We compute all oriented matroids with forbidden circuits formed by a triangle of the above list together with edges of the list not containing a vertex of that particular triangle. If there were no such oriented matroid, we would have the result that there is no embedding of the given torus in euclidean 3-space.

There are many of such oriented matroids $\mathcal{X}$ compatible with the complex which already indicates that there might exist such an embedding. We use the following oriented matroid to define our matroid polyhedron.

+	1234	+	1235	+	1236	+	1237	−	1238
−	1245	−	1246	−	1247	+	1248	−	1256
−	1257	+	1258	−	1267	+	1268	+	1278
+	1345	+	1346	+	1347	−	1348	−	1356
−	1357	−	1358	−	1367	−	1368	−	1378
+	1456	+	1457	−	1458	+	1467	−	1468
+	1478	−	1567	+	1568	+	1578	+	1678
−	2345	−	2346	−	2347	−	2348	−	2356
−	2357	−	2358	−	2367	−	2368	−	2378
−	2456	−	2457	+	2458	−	2467	+	2468
+	2478	+	2567	+	2568	+	2578	+	2678
−	3456	−	3457	−	3458	−	3467	−	3468
−	3478	+	3567	+	3568	−	3578	−	3678
+	4567	−	4568	−	4578	−	4678	+	5678

This oriented matroid has been depicted by providing all rank 3 contractions in Fig. 13. When we compare its corresponding reorientation class with all 24 non-representable reorientation classes from the classification in [16], we find it to be realizable. There is only one non-representable example with 8 elements in rank 4 with 13 simplicial cells, and in this case, the structure of the simplicial cells is not isomorphic with the above case. The result of the classification, summarized by providing all non-realizable reorientation classes of oriented matroids with 8 points in rank 4, was presented in Section 5.

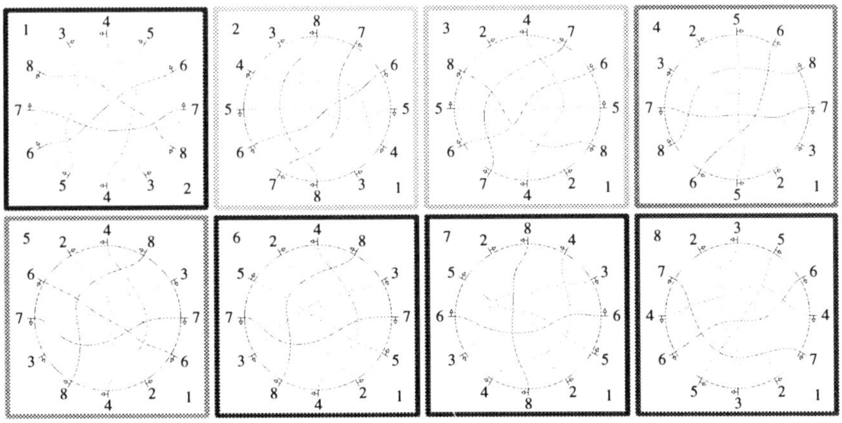

Fig. 13. Oriented matroid for Sarkaria's example of a polyhedron.

We come back to our matroid polyhedron $(\mathcal{X}, \mathcal{C})$. We can store the information about the oriented matroid in terms of its cocircuits.

(	8	1, 2, 3	4, 5, 6, 7	)	(	3, 5, 6, 7	1, 2, 4	8	)
(	3, 6, 7	1, 2, 5	4, 8	)	(	3, 7	1, 2, 6	4, 5, 8	)
(	3	1, 2, 7	4, 5, 6, 8	)	(	4, 5, 6, 7	1, 2, 8	3	)
(	8	1, 3, 4	2, 5, 6, 7	)	(	4, 6, 7, 8	1, 3, 5	2	)
(	4, 7, 8	1, 3, 6	2, 5	)	(	4, 8	1, 3, 7	2, 5, 6	)
(	2	1, 3, 8	4, 5, 6, 7	)	(	2, 8	1, 4, 5	3, 6, 7	)
(	2, 5, 8	1, 4, 6	3, 7	)	(	2, 5, 6	1, 4, 7	3, 8	)
(	3, 7	1, 4, 8	2, 5, 6	)	(	2, 3, 7	1, 5, 6	4, 8	)
(	2, 3	1, 5, 7	4, 6, 8	)	(	3, 4, 6, 7	1, 5, 8	2	)
(	2, 3, 5	1, 6, 7	4, 8	)	(	3, 4, 7	1, 6, 8	2, 5	)
(	3	1, 7, 8	2, 4, 5, 6	)	(	1, 5, 6, 7, 8	2, 3, 4		)
(	1, 6, 7, 8	2, 3, 5	4	)	(	1, 7, 8	2, 3, 6	4, 5	)
(	1, 8	2, 3, 7	4, 5, 6	)	(		2, 3, 8	1, 4, 5, 6, 7	)
(	3, 6, 7	2, 4, 5	1, 8	)	(	3, 7	2, 4, 6	1, 5, 8	)
(	3	2, 4, 7	1, 5, 6, 8	)	(	1, 3, 5, 6, 7	2, 4, 8		)
(	3, 4	2, 5, 6	1, 7, 8	)	(	3, 4, 6	2, 5, 7	1, 8	)
(	1, 3, 6, 7	2, 5, 8	4	)	(	3, 4	2, 6, 7	1, 5, 8	)
(	1, 3, 7	2, 6, 8	4, 5	)	(	1, 3	2, 7, 8	4, 5, 6	)
(	1, 6, 7, 8	3, 4, 5	2	)	(	1, 7, 8	3, 4, 6	2, 5	)
(	1, 8	3, 4, 7	2, 5, 6	)	(		3, 4, 8	1, 2, 5, 6, 7	)
(	4	3, 5, 6	1, 2, 7, 8	)	(	4, 6, 8	3, 5, 7	1, 2	)
(	4, 6	3, 5, 8	1, 2, 7	)	(	4, 8	3, 6, 7	1, 2, 5	)
(	4	3, 6, 8	1, 2, 5, 7	)	(	4, 5, 6	3, 7, 8	1, 2	)
(	1, 8	4, 5, 6	2, 3, 7	)	(	1, 6, 8	4, 5, 7	2, 3	)
(	2	4, 5, 8	1, 3, 6, 7	)	(	1, 8	4, 6, 7	2, 3, 5	)
(	2, 5	4, 6, 8	1, 3, 7	)	(	1, 2, 5, 6	4, 7, 8	3	)
(	2, 3, 4	5, 6, 7	1, 8	)	(	1, 2, 3, 7	5, 6, 8	4	)
(	1, 2	5, 7, 8	3, 4, 6	)	(	1, 2, 5	6, 7, 8	3, 4	)

The circuits (Radon-partitions) are provided in the following list.

(	6	1,2,3	4,5,7,8	)	(	3,5,8	1,2,4	6,7	)
(	3,6,8	1,2,5	4,7	)	(	4,7	1,2,6	3,5,8	)
(	3,4,5,8	1,2,7	6	)	(	3,6	1,2,8	4,5,7	)
(	5,7	1,3,4	2,6,8	)	(	6	1,3,5	2,4,7,8	)
(	2,4,7,8	1,3,6	5	)	(	5	1,3,7	2,4,6,8	)
(	4,5,7	1,3,8	2,6	)	(	2,3,6,8	1,4,5	7	)
(	7	1,4,6	2,3,5,8	)	(	3,5,8	1,4,7	2,6	)
(	2,6	1,4,8	3,5,7	)	(	2,3,4,8	1,5,6	7	)
(	6	1,5,7	2,3,4,8	)	(	2,4,7	1,5,8	3,6	)
(	2,3,4,8	1,6,7	5	)	(	3,5	1,6,8	2,4,7	)
(	2,4,6	1,7,8	3,5	)	(	1,6	2,3,4	5,7,8	)
(	1,4	2,3,5	6,7,8	)	(	5,7,8	2,3,6	1,4	)
(	1,4,5	2,3,7	6,8	)	(	1,4,5,7	2,3,8	6	)
(	7	2,4,5	1,3,6,8	)	(	1,3,5,8	2,4,6	7	)
(	1,6	2,4,7	3,5,8	)	(	5,7	2,4,8	1,3,6	)
(	7,8	2,5,6	1,3,4	)	(	1,4	2,5,7	3,6,8	)
(	7	2,5,8	1,3,4,6	)	(	3,5,8	2,6,7	1,4	)
(	1,3,4,5	2,6,8	7	)	(	6	2,7,8	1,3,4,5	)
(	2,7,8	3,4,5	1,6	)	(	1,5	3,4,6	2,7,8	)
(	2,6,8	3,4,7	1,5	)	(	1,2,6	3,4,8	5,7	)
(	2,7,8	3,5,6	1,4	)	(	1,2,4	3,5,7	6,8	)
(	1,2,4,7	3,5,8	6	)	(	5,8	3,6,7	1,2,4	)
(	5	3,6,8	1,2,4,7	)	(	1,2,4,6	3,7,8	5	)
(	2,3,7,8	4,5,6	1	)	(	1	4,5,7	2,3,6,8	)
(	2,7	4,5,8	1,3,6	)	(	2,3,5,8	4,6,7	1	)
(	1,3,5	4,6,8	2,7	)	(	2,6	4,7,8	1,3,5	)
(	1	5,6,7	2,3,4,8	)	(	7	5,6,8	1,2,3,4	)
(	1,2,3,4	5,7,8	6	)	(	5	6,7,8	1,2,3,4	)

In using the above information, it is not very difficult to find coordinates. But for methods of finding coordinates see the next section.

7. How to find a matrix for an oriented matroid

The problem of finding coordinates for a given oriented matroid is both, a natural and a hard problem (NP-hard). In the general case, we cannot hope for an easy solution. Several heuristic methods are the best tools available, and as long as the problem size is limited, these methods have been tested successfully. There are methods assuming the existence of coordinates and attempts to determine them, and there are methods assuming non-existence of coordinates and attempts to prove this. See Section 5.1, and Table 6 in [9], Section 8 in [5], and the references cited there.

We use the example of a matroid polyhedron we considered earlier and we show the method how to find coordinates in this case. The oriented matroid $\mathcal{X}$ is given, and we look for a 8×4-matrix of representing 8 points with its homogeneous coordinates such that they represent the given oriented matroid. We consider the coordinates as variables which have to obey algebraic inequalities. Looking for a solution of these inequalities is our objective. We analize these inequalities, and we find the system to be very much redundant. A 3-summand Grassmann Plücker relation has in the uniform case always a single summand with a sign opposite to the remaining summands. The sign of a factor within such a single summand must have its correct sign whenever the remaining 5 determinants have their proper signs. Otherwise the Grassmann Plücker relation is violated. In other words: certain inequalities can be discarded without losing the full generality of the solution set.

This idea can be used in a strict manner to reduce the inequality system as much as possible. In our example we have at the beginning a system of 70 inequalities. When introducing a unit matrix for a certain set of 4 points, this comes down to the following types of determinants.

1 times a 4×4-determinant,
16 times a 3×3-determinant,
36 times a 2×2-determinant,
16 times a 1×1-determinant,
1 times a 0×0-determinant.

Any non-reducible inequality system must contain at least the set of inequalities corresponding to mutations, because these signs can never be determined by the remaining signs. Calculating all these mutations leads to altogether 13 of them:

1238, 1245, 1278, 1347, 1367, 1468, 1478, 2348, 2356, 2567, 3568, 3578, 4567

A good choice for the unit matrix is the one consisting of the rows 1,3,5,7, and we get a matrix of the form

$$\begin{array}{c} 1\\2\\3\\4\\5\\6\\7\\8 \end{array}\begin{pmatrix} 1 & 0 & 0 & 0 \\ a=[2357] & b=[1257] & c=[1327] & d=[1352] \\ 0 & 1 & 0 & 0 \\ e=[4357] & f=[1457] & g=[1347] & h=[1354] \\ 0 & 0 & 1 & 0 \\ i=[6357] & j=[1657] & k=[1367] & l=[1356] \\ 0 & 0 & 0 & 1 \\ m=[8357] & n=[1857] & o=[1387] & p=[1358] \end{pmatrix}$$

From the entries $a, b, \ldots, p$ of the submatrix of the rows 2,4,6,8, the so called Tucker matrix, we can calculate all the remaining determinants. Considering all the mutations, we have to solve at least the following system of inequalities corresponding to the following determinants

twice a 3×3-determinant:	1468, 2348
8 times a 2×2-determinant:	1238, 1245, 1278, 1478,
	2356, 2567, 3568, 4567

with the variables $a, b, c, \ldots, p$ of known signs. On the other side, it turns out in this case that all the other signs are automatically fulfilled whenever the above inequalities together with the inequalities requiring the signs of the variables are fulfilled. A transition to the absolute values of the 16 variables leads to a small algebraic non-linear inequality system for 16 positiv numbers. Now the remaining task of solving this reduced system is no longer very difficult and can be done by hand.

Acknoledgement. I wish to express my gratitude to all my students and former students who made contributions to implementations for handling oriented matroids. From the very early versions mentioned in the Springer Lecture Notes Volume [22] with B. Sturmfels, there exist now many additional implementations. The OMAWIN-program was supported in part by the Deutsche Forschungsgemeinschaft (DFG). Its early version was develloped with J. Richter-Gebert in connection with several open problems in the theory of oriented matroids. In the meantime it was extended very much to its present version by P. Schuchert, M. Gebhard and D. Klodt. Additional implementations by A. Guedes de Oliveira, T. K. Strempel, K. P. Pock, J. Scharnbacher, U. Thiemann and C. Engel were essential contributions for working with oriented matroids as well. RADONCHI and CHIMAN written by P. Schuchert are other very powerful tools for generating oriented matroids with special properties and for handling oriented matroids.

References

1. Altshuler, A., Bokowski, J. and Schuchert, P.: 1995, Neighborly 2-manifolds with 12 vertices, *Manuscript.*
2. Altshuler, A., Bokowski, J. and Steinberg, L.: 1980, The classification of simplicial 3-spheres with nine vertices into polytopes and nonpolytopes, *Discrete Math.,* **31**, pp. 115–124.
3. Bachem, A. and Kern, W.: 1992, Linear Programming Duality. An Introduction to Oriented Matroids, *Universitext*, Springer, Berlin-Heidelberg.
4. Barthel, G., Hirzebruch, F. and Höfer, T.: 1987, Geradenkonfigurationen und Algebraische Flächen, *Vieweg Verlag.*
5. Björner, A., Las Vergnas, M., Sturmfels, B., White, N. and Ziegler, G.M.: 1993, Oriented matroids, *Cambridge University Press, Cambridge.*
6. Bohne, J.: 1992, Eine kombinatorische Analyse zonotopaler Raumaufteilungen, *Ph.D. Thesis*, Bielefeld.
7. Bohne, J., Dress, A. W. M. and Fischer, S.: 1989, A simple proof of de Bruijn's dualization principle. In: Some elementary proofs for some elementary facts in algebra and geometry, *preprint 89-127, SFB 343, Bielefeld.*
8. Bokowski, J.: 1991, On the geometric flat embedding of abstract complexes with symmetries. In: Symmetry of discrete mathematical structures and their symmetry groups. A collection of essays; Hofmann, K. H. and Wille, R. (eds.), *Research and Exposition in Mathematics*, 15, Heldermann, Berlin, pp. 1–48.
9. Bokowski, J.: 1993, Oriented matroids, *Handbook of Convex Geometry, Gruber P. and Wills J.M., (eds.), Elsevier, North-Holland, Netherland.*
10. Bokowski, J.: 1994, On recent progress in computational synthetic geometry, in POLYTOPES: abstract, convex and computational, eds. Bisztriczky, T., McMullen, P., Schneider, R. and Ivić Weiss, A., *Kluwer Academic Publishers, Series C: Mathematical and Physical Sciences* **440**, pp.335–358.
11. Bokowski, J.: 1995, Oriented matroid theory and its applications. Selected concepts and methods, in *Proceedings of Conference on Integral Geometry, Geometrical Probability, and Convex bodies, Cagliari, Sardinia, Italy..*
12. Bokowski, J. and Guedes de Oliveira, A.: 1994, Invariant theory-like theorems for matroids and oriented matroids, *Advances in Math.* **109,1**, pp. 34–44.
13. Bokowski, J., Guedes de Oliveira, A. and Richter-Gebert, J.: 1991, Algebraic varieties characterizing matroids and oriented matroids, *Advances in Math.* **87,2**, pp. 160–185.
14. Bokowski, J., Guedes de Oliveira, A., Thiemann, U. and Veloso da Costa, A.: 1995, On the cube problem of Las Vergnas, *Geometriae Dedicata*, to appear.
15. Bokowski, J., Laffaille, G. and Richter, J.: 1991, Classification of nonstrechable pseudoline arrangements and related properties, *Manuscript.*

16. Bokowski, J. and Richter-Gebert, J.: 1991, On the classification of non-realizable oriented matroids, *Manuscript*, 16p.
17. Bokowski, J. and Schuchert, P.: 1995, Altshuler's sphere M^9_{963} revisited, *SIAM J. on Discrete Mathematics,* to appear.
18. Bokowski, J. and Schuchert, P.: 1995, Equifacetted 3-spheres as topes of nonpolytopal matroid polytopes, *Discrete Comput. Geometry,* **13**, pp. 347–361.
19. Bokowski, J. and Strempel, T. K.: 1995, Pseudogeraden Arrangements mit maximaler Dreiecksanzahl, *Manuscript.*
20. Bokowski, J. and Sturmfels, B.: 1985, Programmsystem zur Realisierung orientierter Matroide, *Preprints in Optimization,* PD 85.22, pp. 1–33.
21. Bokowski, J. and Sturmfels, B.: 1989, An infinite family of minor-minimal nonrealizable 3-chirotopes, *Math. Z.*, **200**, pp. 583–589.
22. Bokowski, J. and Sturmfels, B.: 1989, Computational synthetic geometry, *Lecture Notes in Mathematics*, 1355, Springer, Heidelberg.
23. Cervone, D. P.: 1993, Vertex-minimal immersions of simplicial surfaces into three-space, *Ph.D. Thesis, Brown University, Rhode Island.*
24. Cervone, D. P.: 1993, Vertex-minimal immersions of the Klein bottle, *Geometriae Dedicata,* to appear.
25. Folkman, J. and Lawrence, J.: 1978, Oriented matroids, *J. Comb. Theory, Ser. B,* **25**, pp. 199–236.
26. Goodman, J. E. and Pollack, R.: 1984, Semispaces of configurations, cell complexes of arrangements, *Journal of Combinatorial Theory, Ser. A*, **37**, pp. 257–293.
27. Grünbaum B.: 1972, Arrangements and Spreads, *Regional Conf., Vol. 10 (Amer. Math. Soc., Providence, RI).*
28. Harborth, H.: 1985, Some simple arrangements of pseudolines with a maximum number of triangles, *Disc. Geom. Conv., N.Y. Acad. Sc.* **440**, pp. 30–31.
29. Harborth, H.: 1988, Problem about a simple arrangement of pseudolines, *Math. Semesterberichte.*
30. Las Vergnas, M.: 1975, Matroïdes orientables, *C.R. Acad. Sci. Paris, Ser. A,***280**, pp. 61–64.
31. Las Vergnas, M.: 1980, Convexity in oriented matroids, *J. Comb. Theory, Ser. B,***29**, pp. 231–243.
32. Las Vergnas, M., Roudneff, J.-P. and Salaün, I.: 1991, Regular polytopes and oriented matroids, *preprint.*
33. Levi, F.: 1926, Die Teilung der projektiven Ebene durch Gerade und Pseudogerade, *Ber. Math.-Phys. Kl. Sächs. Akad. Wiss.,* **78**, pp. 256–267.
34. Ljubić, D., Roudneff, J.-P. and Sturmfels, B.: 1989, Arrangements of lines and pseudolines without adjacent triangles, *J.Comb.Theor.Ser.A.,* **50**, pp. 24–32.
35. Mnëv, N.E., The universality theorems on the classification problem of configuration varieties and convex polytopes varieties, in: Rohlin Seminar, ed. O.Y. Viro, *Lecture Notes in Mathematics,* **1346**, Springer, Heidelberg, pp. 1–13.
36. Richter–Gebert, J.: 1992, Oriented matroids with few mutations, *Discrete Comp. Geom.* **10**, pp. 251–269.
37. Richter–Gebert, J.: 1992, On the realizability problem of combinatorial geometries – Decision methods, *Ph.D. Thesis, University Darmstadt.*
38. Salaün, I.: 1988, Deux problèmes de gèometrie combinatroire, *Thèse de troisième Cycle, Université Paris 6.*
39. Schuchert, P.: 1995, Matroid-Polytope und Einbettungen kombinatorischer Mannigfaltigkeiten, *Ph.D. Thesis, University Darmstadt.*
40. Shannon, R. W.: 1979, Simplicial cells in arrangements of hyperplanes, *Geometriae Dedicata,* **8**, pp. 179–187.

A two-element contraction of the 600-cell.

Vertex figure of the dodecahedron.

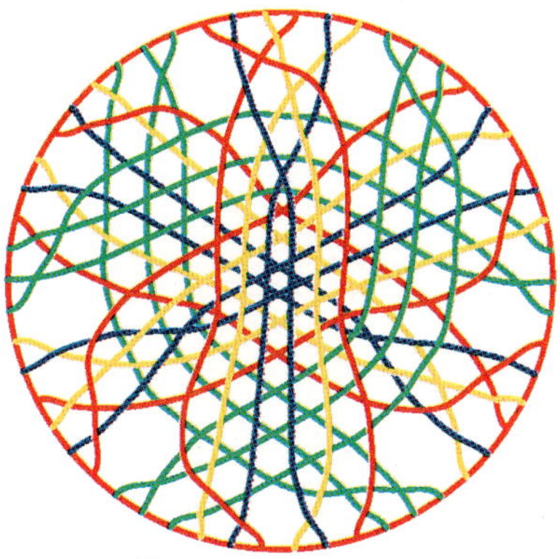

Harborth's arrangement.

Corresponding zonotopal tiling.

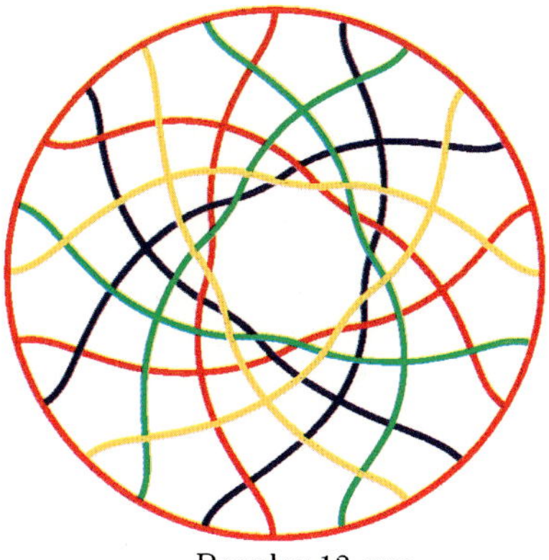

Regular 12-gon.

Corresponding zonotopal tiling.

Fig. 14. Examples in rank 3

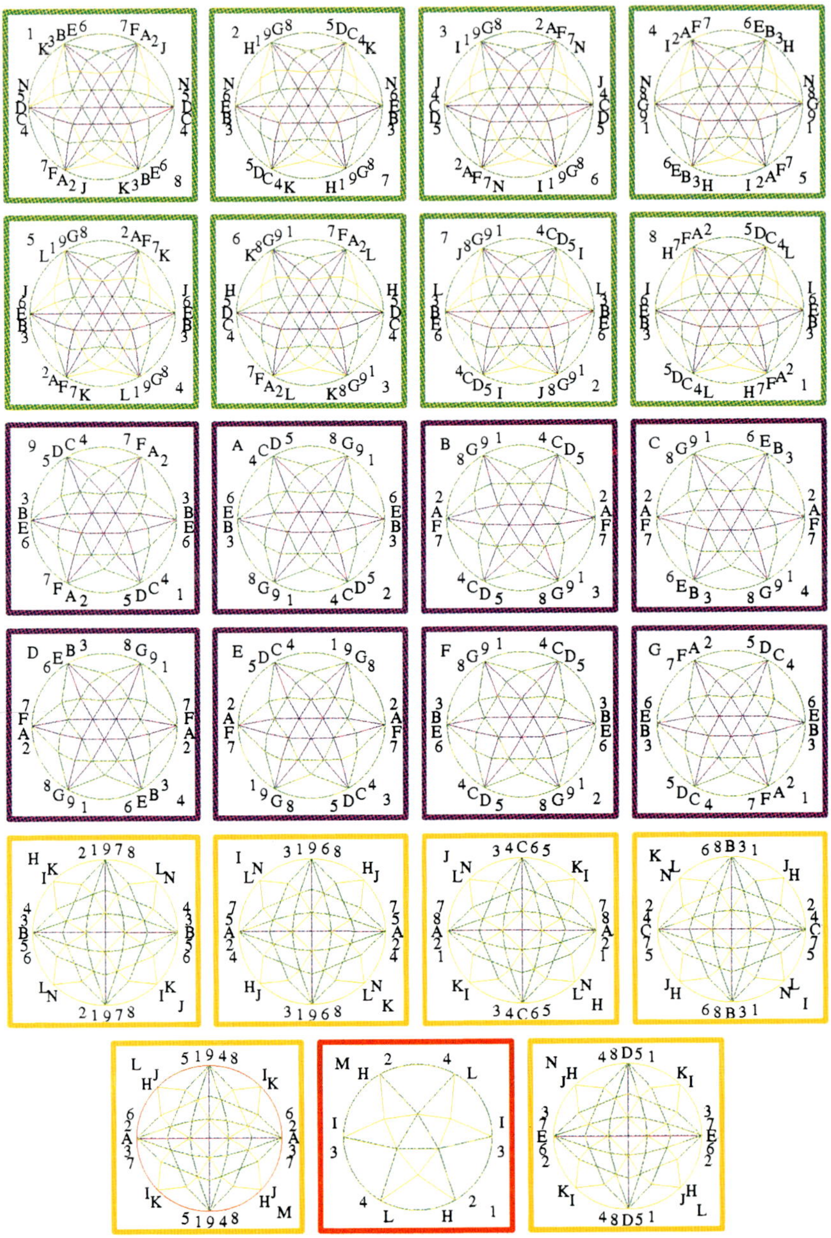

Fig. 15. All 23 contractions of the vertex figure of the 24-cell.

A Survey of Geometric Reasoning Using Algebraic Methods*

Shang-Ching Chou and Xiao-Shan Gao†

Department of Computer Science
The Wichita State University, Wichita, KS 67260, USA
e-mail: (chou,gao)@.cs.twsu.edu

Abstract. In the past decade highly successful algebraic methods for mechanical geometry theorem proving have been developed. The first step in these methods is to assign coordinates to key points and to translate the hypotheses and conclusion of a geometry statement into multivariate polynomial equations and inequalities. The second step is to prove the corresponding algebraic statements using various algebraic techniques. To date the most practically successful algebraic techniques have been Ritt–Wu's characteristic set (CS) method and the Gröbner basis (GB) method. Also Collins' method, a quantifier elimination method for real closed fields of Tarski's type, has been practically improved to such an extent that many non-trivial geometry problems can now be solved by computer programs based on this method.

This survey mainly concentrates on the applications of the characteristic set method (the CS method) and the Gröbner basis method (the GB method) to automated reasoning in elementary geometries, differential geometries, and mechanics. We will use elementary and understandable examples to show the essence of the techniques we are using, letting the reader to consult the related references for more detailed issues underlying these techniques.

Keywords. Theorem proving, formula derivation, elementary geometry, differential geometry, characteristic set method, Gröbner bases, Tarski's decision procedure, Collins' method, algebraic variety, decomposition of algebraic set, Simson's theorem, Morley's trisector theorem, Kepler–Newton problem, Heron's formula, Peaucellier's Linkage.

1. A History Review

Generally, there are two approaches to proving theorems using computers: the human simulation approach and the logic approach. The earliest work in geometry theorem proving by computer programs was done by H. Gelernter and his collaborators [27]. It was based on the human simulation approach and has been considered a landmark in the AI area for its time. L. A. Wos and his collaborators used their powerful general-purpose resolution theorem prover (based on the logic approach) to experiment with proving theorems in Tarski's axioms for elementary geometry [42]. In spite of the success with these techniques, the results did not lead to the development of an automated *and powerful* geometry prover.

Algebraic approaches for proving geometry theorems can be considered a special kind of logic approach. In this area, the earliest work dates back to Hilbert. In his classic book [29], Hilbert outlined a decision method for constructive geometry statements in *affine geometry*. As Tarski pointed out, Hilbert's result "is closely connected with the decision method for elementary geometry, but has a rather restricted character" [47].

In 1948, Tarski published a decision method for the theory of real closed fields, thus giving a decision method for what he called elementary geometry [47]. In spite of subsequent improvements by A. Seidenberg [46] and others, for years variations of Tarski's method remained

* The work reported here was supported in part by the NSF Grant CCR-917870.

† On leave from Institute of Systems Science, Academia Sinica, Beijing.

impractical for proving non-trivial theorems in geometry.

In 1974, G. Collins made an important contribution along the Tarski line [22]. His cylindrical algebra decomposition (CAD) algorithm is currently the best of the Tarski type algorithms. This method was implemented by D. Arnon, and several difficult algebra-geometry related problems were solved by Arnon's program [1] [2]. More recently, another major practical improvement of Collins' method has been made by Collins and H. Hong [23]. Thanks to the generosity of Collins and Hong, we have been able to experiment with proving geometry theorems using Hong's program. The results were very encouraging. We use the following two examples to give some idea about how far Collins' method can reach now.

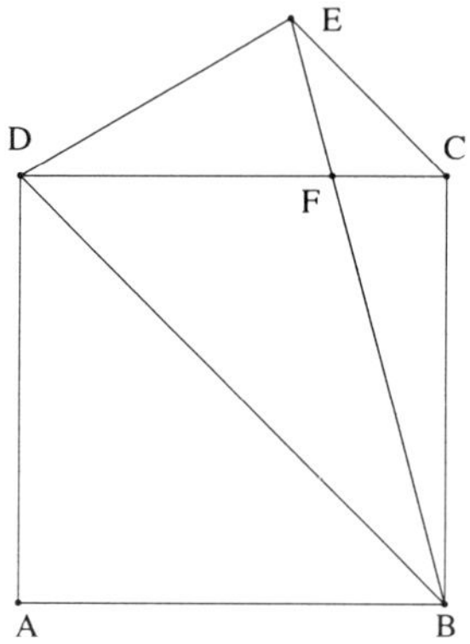

Fig. 1

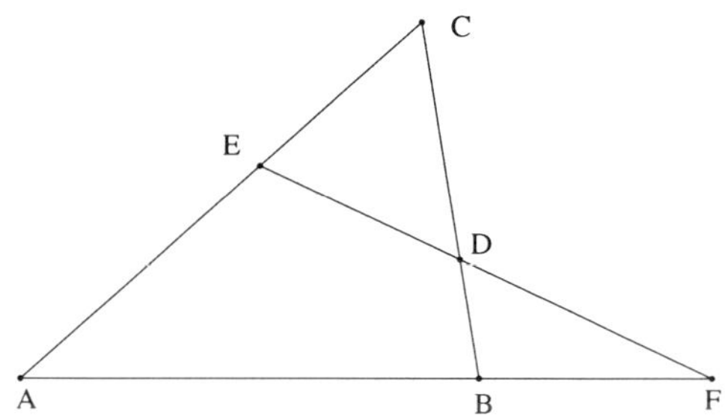

Fig. 2

Example (1.1). Let $ABCD$ be a square. CG is parallel to the diagonal BD. Point E is on CG such that $BE = BD$. F is the intersection of BE and DC. Show that $DF = DE$ (Fig. 1).

Let $A = (0,0)$, $B = (u_1, 0)$, $C = (u_1, u_1)$, $D = (0, u_1)$, $E = (x_1, x_2)$ and $F = (x_3, u_1)$. Then the hypotheses can be expressed by the following polynomial equations:

$$h_1 = x_2^2 + x_1^2 - 2u_1x_1 - u_1^2 = 0 \qquad BE = BD$$
$$h_2 = u_1x_2 + u_1x_1 - 2u_1^2 = 0 \qquad CE \parallel BD$$
$$h_3 = x_2x_3 - u_1x_2 - u_1x_1 + u_1^2 = 0 \qquad F \text{ is on } BE.$$

The conclusion ($DF = DE$) can be expressed by $c = (x_3-0)^2+(u_1-u_1)^2-[(x_1-0)^2+(x_2-u_1)^2] = x_3^2 - x_2^2 + 2u_1x_2 - x_1^2 - u_1^2 = 0$.

Thus the algebraic form of the above statement is:

$$\forall u_1x_1x_2x_3[(h_1 = 0 \wedge h_2 = 0 \wedge h_3 = 0 \wedge u_1 \neq 0) \Rightarrow c = 0],$$

which was proved to be valid by Hong's program in 16 seconds on a SPARC Station 1. This theorem has been considered fairly difficult in high school geometry. (Try finding a traditional proof.) Also the previous implementation of Collins' method (Arnon's program) was unable to prove this theorem within reasonable computer resources. A more difficult problem is:

Example (1.2) (Pasch). A line intersects the three sides of a triangle ABC at points D, E, and F. If D is between BC and E is between AC, then F is outside AB (Fig. 2).

Let $A = (0,0)$, $B = (u_1, 0)$, $C = (u_2, u_3)$, $D = (x_2, x_1)$ $E = (x_4, x_3)$, and $F = (x_5, 0)$.

Then the hypotheses can be expressed by

$H_1 : h_1 = u_3x_2 - (u_2 - u_1)x_1 - u_1u_3 = 0$	D is on BC
$H_2 : h_2 = (x_2 - u_1)(u_2 - x_2) > 0$	D is between B and C
$H_3 : h_3 = (x_4 - u_2)(0 - x_4) > 0$	E is between A and C
$H_4 : h_4 = u_3x_4 - u_2x_3 = 0$	E is on AC
$H_5 : h_5 = (x_3 - x_1)x_5 + x_1x_4 - x_2x_3 = 0$	D, E, and F are collinear
$H_6 : h_6 = u_1u_3 \neq 0$	A, B, and C are not collinear

The conclusion that F is outside AB can be expressed by $g = (x_5 - 0)(x_5 - u_1) > 0$.

Thus the algebraic equivalence of the above statement is:

$$\forall u_1 \cdots u_3 x_1 \cdots x_5[(H_1 \wedge \cdots \wedge H_6) \Rightarrow g > 0],$$

which was proved to be valid by Hong's program in 2,156 seconds.

However, Collins' method requires further improvements in order to prove a substantial number of non-trivial theorems in practice. On the other hand, it is theoretically capable of addressing a much wider range of problems than the CS and GB methods. It can address statements involving not only equations, but also inequalities; not only statements with all universal (or all existential) quantifiers outside, but also statements with universal and existential quantifiers in any places.

Practically, P.J. Davis appears to be the first to explore the algebraic approach to proving geometry theorems using the computer [24]. His approach for the computer proof of Pappus' theorem is essentially the one described by Hilbert, but he did not provide a unifying mechanical way to do it.

Restricting himself to a class of geometry statements of *equality type*, Wen–Tsün Wu introduced a method in 1977 which could be used to prove quite non-trivial geometry theorems [49]. Wu's approach can be considered a generalization of Hilbert's approach. Wu's work became known outside China mainly through the papers [51] [7], and the fact that over 130 theorems were proved by the method in [7] was quite encouraging. H. Ko and M. Hussain [34] also succeeded in implementing a theorem prover based on [7]. Later it was clarified [50] that the algebraic tools needed in Wu's approach had first been developed by J. Ritt [44] [45]. The algebraic aspect of this approach is now known as the characteristic set (CS) method. We use Wu's method to refer to his approach for geometry theorem proving.

The success of Wu's method has revived interest in proving geometry theorems by computers. In particular, the application of the Gröbner basis (GB) method, long known in computer algebra community, to the same class of geometry theorems that Wu's method addresses has been investigated. In 1985–1986, three groups ([20], [31], and [37]) reported practical successes. It is now the case that hundreds of theorems in Euclidean and non-Euclidean geometries can be proved automatically by computer programs based on these two methods [10] [40] [25] [26]. Here we would like to remind the reader that Wu's method and the GB method can only deal with theorems involving equalities, but not inequalities. They cannot address, e.g., Example (1.2).

Recently, methods of proving geometry theorems by checking examples based on the CS method have been proposed [30] [63] [62]. An implementation of the method in [62] is also very successful.

At the same time, automated derivation of geometry locus equations and other geometry formulas using the CS and GB methods were investigated [7] [8] [55] [9]. Other variations for dealing with such kind problems have been reported in [12]. About 120 problems in geometry were solved by these methods in [13].

All the above work is concerned with *elementary geometries* in Wu's sense, i.e., geometries in which no differentiation is involved. The CS method originally developed by Ritt is also applicable to *differential polynomials.* In another pioneering paper [58], Wu extended the CS method to prove theorems in *differential geometry.* While extensive computer experiments with this method started only recently [56] [18], the results are encouraging, and nearly 100 non-trivial theorems in space curve theory have been proved.

There have also been several successful applications of the CS method to mechanics [57] [19]; notably, automated proofs of Newton's laws from Kepler's laws were given. In these works, computer experiments in automated formula derivation in differential geometry and mechanics were also discussed.

This survey summarizes the results in automated reasoning in geometries using the CS and GB methods.

The GB method was first introduced by B. Buchberger [4]. It is an important tool for solving problems in the polynomial ideal theory. In particular, it solves the ideal membership problem. The reader may consult [5] for details.

The CS method was introduced by Ritt in connection with the study of the algebraic properties of a system of differential equations [44] [45]. Its algorithmic aspects have been studied seriously since Wu's work [50]. The method provides a decomposition of an algebraic set into the union of irreducible varieties. In particular, it solves the radical ideal membership problem. A method for computing a projection of an algebraic set has been proposed based on the CS method [60]. This leads to a CS method for the quantifier elimination for the theory of algebraically closed fields. There are no complete GB methods in the cases of differential polynomials. In particular, the decision problem of ideal membership remains open for differential polynomials. For a detailed presentation of the CS method see [50] [10] [11].

2. Theorem Proving in Elementary Geometries

This is the most developed and successful area using the CS and GB methods. Roughly speaking, the two methods can address those geometry statements, for which, in their algebraic form, the hypotheses can be expressed by a set (conjunction) of polynomial equations

$$\begin{array}{ll} & h_1(y_1,...,y_m) = 0 \\ & h_2(y_1,...,y_m) = 0 \\ (2.1) & \quad\cdots \\ & h_r(y_1,...,y_m) = 0, \end{array}$$

and the conclusion is also a polynomial equation $c(y_1,...,y_m) = 0$, where the h's and c are polynomials with coefficients in a *base field* K. Usually, as we shall assume in this paper, $K = \mathbf{Q}$, the field of rational numbers. Thus the algebraic form of the geometry statement would be

$$\forall y[(h_1 = 0 \wedge \cdots \wedge h_r = 0) \Rightarrow c = 0].$$

However, such formulas are usually not valid because most geometry statements are only valid under some non-degenerate conditions. Let us look at two concrete examples to see the real situations.

2.1. Two Examples and the Basic Method

Example (2.1). Let $ABCD$ be a parallelogram, and E be the intersection of the two diagonals AC and BD. Show that $AE = CE$ (Fig. 3).

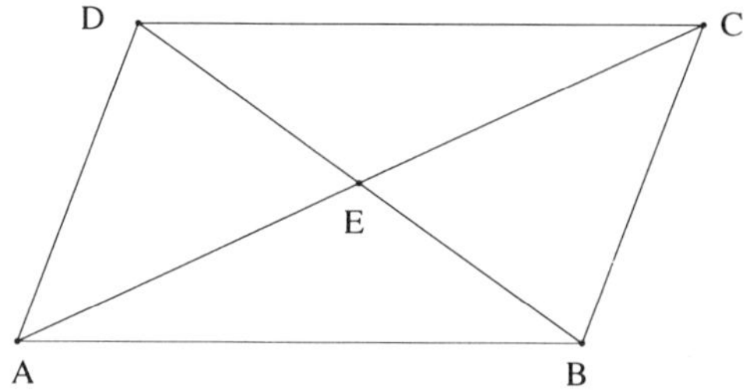

Fig. 3

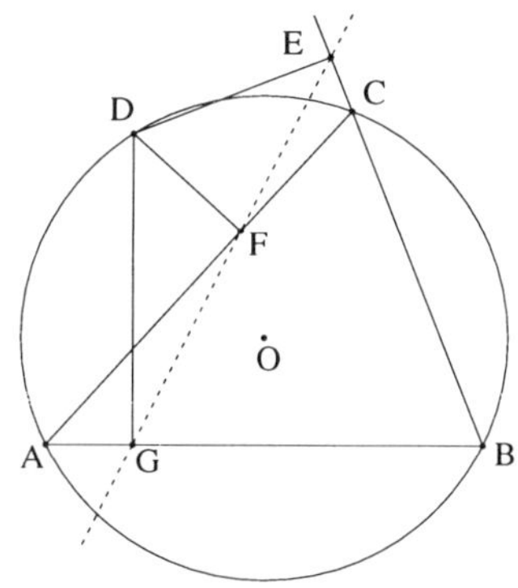

Fig. 4

The hypotheses of the geometric conditions are $DC \parallel AB$, $DA \parallel BC$, E is on AC, and E is on BD.

The first step is to assign coordinates to the points, and then translate the hypotheses and conclusion of the statement into polynomial equations.

We can choose point A as the origin, and line AB as the x–axis of the coordinate system: $A = (0, 0)$, $B = (u_1, 0)$. We can assign the coordinates to point $C = (u_3, u_2)$. Since points A, B, and C can be arbitrarily chosen, their non-zero coordinates are independent variables or parameters, denoted by u's. Once these three points are fixed, the other points D and E are determined; their coordinates are dependent variables, denoted by x's. We can assign coordinates to D and E simply by increasing dependent variable indexes: $E = (x_2, x_1)$, and $D = (x_4, x_3)$. The selection of the parameters u's and the dependent variables x's and the order of dependent variables can be done automatically by computers, at least for a class of geometry statements of constructive type, including the two examples in this section (see [8] or [17]).

Once we have the coordinate assignment, the conversion of the above hypotheses and conclusion to their algebraic forms is straightforward and can be done also mechanically. Thus we have four equations corresponding to the above four geometric conditions:

$$
\begin{array}{llr}
 & h_1 = u_1 x_1 - u_1 u_2 = 0 & DC \parallel AB \\
 & h_2 = u_2 x_2 - (u_3 - u_1) x_1 = 0 & DA \parallel BC \\
(2.2) & h_3 = u_2 x_4 - u_3 x_3 = 0 & E \text{ is on } AC \\
 & h_4 = x_1 x_4 - (x_2 - u_1) x_3 - u_1 x_1 = 0 & E \text{ is on } BD.
\end{array}
$$

The conclusion that $AE = CE$ can be expressed by $c = x_4^2 + x_3^2 - [(x_4 - u_3)^2 + (x_3 - u_2)^2] = 2u_3 x_4 + 2u_2 x_3 - u_3^2 - u_2^2 = 0$.

Thus the tentative algebraic form of our parallelogram theorem would be

$$\forall ux[(h_1 = 0 \wedge h_2 = 0 \wedge h_3 = 0 \wedge h_4 = 0) \Rightarrow c = 0].$$

The above formula is "almost" correct except for a non-degenerate condition: it is not valid when A, B, and C are collinear.

A nice feature of the method introduced by Wu is that non-degenerate conditions sufficient for a geometry statement to be valid can be generated automatically during the proof process. Let us show how Wu's method works. The basic operation of the method is pseudo division.

Pseudo Division. Let

$$
\begin{array}{ll}
g = a_n y^n + \cdots + a_1 y + a_0 & (a_n \neq 0) \\
f = b_k y^k + \cdots + b_1 y + b_0 & (b_k \neq 0 \wedge k > 0)
\end{array}
$$

be two polynomials in the variable y, where the coefficients a_i, b_i are in $\mathbf{Q}[v_1, ..., v_d]$, $y \notin \{v_1, ..., v_d\}$. Here a_n (or b_k) and n (or k) are called the leading coefficient and leading degree of g (or f) in the variable y and denoted by $lc(g, y)$ and $ld(g, y)$ (or $lc(f, y)$ and $ld(f, y)$), respectively. We want to divide g by f in the variable y. If b_k is 1, then we use long division from high schools; otherwise we use pseudo division, which is similar to the long division. For our purpose, all we really care about is the pseudo remainder r, denoted by $r = prem(g, f, y)$. The details of this process are covered in the Appendix, but for now the important property of this algorithm is captured in the following remainder formula:

$$b_k^s \cdot g = q \cdot f + r, \qquad ld(r, y) < k, \tag{2.3}$$

where s $(\leq n - k + 1)$ is a non-negative integer. If $b_k = 1$, the above formula is the remainder formula for the ordinary long division.

The next step is to transform the set of hypotheses into a triangular form, where each equation introduces only one new dependent variable with a higher index. Let us look at (2.2); h_1 introduces x_1 and h_2 introduces x_2; so far so good. But h_3 introduces two new dependent variables x_3 and x_4 at the same time; thus (2.2) is not in triangular form. However, it is very easy to transform it into a triangular form by letting $f_1 = h_1$, $f_2 = h_2$, $f_3 = prem(h_4, h_3, x_4)$, $f_4 = h_4$. Then $f_1, ..., f_4$ is a triangular form or an ascending chain ASC:[1]

$$\begin{aligned} f_1 &= u_1x_1 - u_1u_2 = 0 \\ f_2 &= u_2x_2 - (u_3 - u_1)x_1 = 0 \\ f_3 &= (-u_2x_2 + u_3x_1 + u_1u_2)x_3 - u_1u_2x_1 = 0 \\ f_4 &= x_1x_4 - (x_2 - u_1)x_3 - u_1x_1 = 0. \end{aligned} \tag{2.4}$$

Once we have the triangular form, then we can perform the key step of the method: the successive (pseudo) division of c, the conclusion polynomial, with respect to that triangular form, i.e.,

$$R_4 = prem(c, f_4, x_4) = (2u_3x_2 + 2u_2x_1 - 2u_1u_3)x_3 - (u_3^2 - 2u_1u_3 + u_2^2)x_1$$

$$R_3 = prem(R_4, f_3, x_3) = (u_2u_3^2 + u_2^3)x_1x_2 - (u_3^3 - 2u_1u_3^2 + u_2^2u_3 - 2u_1u_2^2)x_1^2 - (u_1u_2u_3^2 + u_1u_2^3)x_1$$

$$R_2 = prem(R_3, f_2, x_2) = (u_1u_2u_3^2 + u_1u_2^3)x_1^2 - (u_1u_2^2u_3^2 + u_1u_2^4)x_1$$

$$R_1 = prem(R_2, f_1, x_1) = 0.$$

The last remainder R_1 is called the final remainder of c with respect to the triangular form $f_1, ..., f_4$; we denote it by $prem(c; f_1, ..., f_4)$ or $prem(c; ASC)$. In this particular case it turns out to be zero by our computation. That means we have proved the theorem. To see this, first we always have the following *remainder formula* when doing a pseudo successive division of any polynomial g with respect to a triangular form $ASC = f_1, ..., f_p$,

$$I_1^{s_1} \cdots I_p^{s_p} g = Q_1 f_1 + \cdots + Q_p f_p + R_1 \tag{2.5}$$

Here the I_k are leading coefficients of the f_k and $s_k \geq 0$ $(k = 1, ..., p)$; $R_1 = prem(c; ASC)$ is the final remainder. In our case, $p = 4$ and the I_k are

$$\begin{aligned} I_1 &= u_1 \\ I_2 &= u_2 \\ I_3 &= -u_2x_2 + u_3x_1 + u_1u_2 \\ I_4 &= x_1. \end{aligned}$$

Formula (2.5) is a generalization of the remainder formula (2.3) for pseudo division and can be easily proved by induction. Since in our case $R_1 = 0$ and $f_k = 0$ $(k = 1, ..., 4)$ by the hypotheses,

[1] Strictly speaking, there are other restrictions for a triangular form to be an ascending chain.

the right side of (2.5) is zero. Thus the conclusion polynomial must be zero if we assume all $I_k \neq 0\ (k = 1, .., 4)$.

$I_k \neq 0$ are usually connected with non-degeneracy. Thus the last step is the analysis of subsidiary conditions $I_k \neq 0$. E.g., $I_1 \neq 0$ and $I_2 \neq 0$ mean points A, B, and C are not collinear; $I_3 \neq 0$ means AC and BD have a normal intersection; $I_4 \neq 0$ means D is not on line AB.

Before going to the next example, let us summarize this simple use of Wu's method:

Step 1. Assign coordinates to the points involved, then translate the geometry hypotheses to a set (conjunction) of polynomial equations $h_1 = 0, ..., h_r = 0$; also the conclusion is a polynomial equation $c = 0$. For geometry statements of constructive type, this can be done totally automatically.

Step 2. Transform the set of hypothesis polynomials into a triangular form (an ascending chain) $ASC = f_1, ..., f_p$. For geometry statements of constructive type, since new points are introduced one by one, the new dependent variables are introduced at most two by two, and this transformation is then easy to do locally (involving at most two equations at a time). Generally, we have a complete triangular algorithm presented and referred to as Ritt's principle by Wu [50] (also see [10]).

Step 3. Divide the conclusion polynomial c successively with respect to the triangular form $ASC = f_1, .., f_p$ to obtain the final remainder $R_1 = prem(c; ASC)$. If $R_1 = 0$, then the statement is confirmed under subsidiary conditions $I_k \neq 0$, where the I_k are the leading coefficients of the f_k.

Step 4. Analyze the subsidiary conditions $I_k \neq 0$ which are usually connected with non-degeneracy. Again, for geometry statements of constructive type, this analysis can be done automatically: non-degenerate conditions can be generated mechanically in their geometric form [8] [17].

The experienced reader might notice that for this simple example, we don't have to use this general schema. We can solve the dependent variables $x_1, ..., x_4$ successively in terms of the u's, then substitute the solutions into the conclusion polynomial c to see whether it is identical to zero. This is exactly what Hilbert's mechanical proof method (mentioned in the introduction) for constructive affine geometry statements does. However, Hilbert didn't mention non-degenerate conditions which are important for a geometry statement to be valid; or put it another way, what Hilbert considered a geometry statement to be true is what we consider it to be generally (generically) true (see Formulation F1 below). Moreover, if we cannot solve dependent variables explicitly, then we have to use the above general schema. The following theorem is such an example.

Example (2.2). (Simson's Theorem). Let D be a point on the circumscribed circle (O) of triangle ABC. From D perpendiculars are drawn to the three sides BC, CA and AB. Let E, F and G be the three feet. Show that E, F and G are collinear (Fig. 4).

Step 1. Converting the statement into its algebraic form, we let $A = (0,0)$, $B = (u_1,0)$, $C = (u_3,u_2)$, $O = (x_1,x_2)$, $D = (x_3,u_4)$, $E = (x_5,x_4)$, $F = (x_7,x_6)$, and $G = (x_9,x_8)$. We then have a set of hypothesis equations H:

$$\begin{array}{lll}
 & h_1 = 2u_1x_1 - u_1^2 = 0 & OA = OB \\
 & h_2 = 2u_2x_2 + 2u_3x_1 - u_3^2 - u_2^2 = 0 & OA = OC \\
 & h_3 = x_3^2 - 2x_1x_3 + u_4^2 - 2x_2u_4 = 0 & DO = OA \\
 & h_4 = u_2x_5 - (u_3 - u_1)x_4 - u_1u_2 = 0 & E \text{ is on } BC \\
(2.6) & h_5 = (u_3 - u_1)x_5 + u_2x_4 - (u_3 - u_1)x_3 - u_2u_4 = 0 & ED \perp BC \\
 & h_6 = u_2x_7 - u_3x_6 = 0 & F \text{ is on } AC
\end{array}$$

$$h_7 = u_3x_7 + u_2x_6 - u_3x_3 - u_2u_4 = 0 \qquad FD \perp AC$$
$$h_8 = u_1x_8 = 0 \qquad G \text{ is on } AB$$
$$h_9 = u_1x_9 - u_1x_3 = 0 \qquad GD \perp AB.$$

The conclusion that E, F, and G are collinear, can be expressed by the equation $c = (x_6 - x_4)x_9 - (x_7 - x_5)x_8 + x_4x_7 - x_5x_6 = 0$.

Step 2. Using the pseudo remainder $prem$ computation, we can triangulize $h_1, ..., h_9$ by letting $f_4 = prem(h_5, h_4, x_5)$, $f_6 = prem(h_7, h_6, x_7)$, $f_i = h_i$ for $i \neq 4, 6$. We thus have a triangular form $ASC = f_1, ..., f_9$:

$$\begin{aligned}
f_1 &= 2u_1x_1 - u_1^2 \\
f_2 &= 2u_2x_2 + 2u_3x_1 - u_3^2 - u_2^2 \\
f_3 &= x_3^2 - 2x_1x_3 + u_4^2 - 2x_2u_4 \\
f_4 &= (u_3^2 - 2u_1u_3 + u_2^2 + u_1^2)x_4 - (u_2u_3 - u_1u_2)x_3 - u_2^2u_4 + u_1u_2u_3 - u_1^2u_2 \\
f_5 &= u_2x_5 - (u_3 - u_1)x_4 - u_1u_2 \\
f_6 &= (u_3^2 + u_2^2)x_6 - u_2u_3x_3 - u_2^2u_4 \\
f_7 &= u_2x_7 - u_3x_6 \\
f_8 &= u_1x_8 \\
f_9 &= u_1x_9 - u_1x_3.
\end{aligned} \tag{2.7}$$

Step 3. Now use successive pseudo division to compute the final remainder $R_1 = prem(c; ASC)$. That is,

$R_9 = prem(R_{10}, f_9, x_9) = -(u_1x_7 - u_1x_5)x_8 + u_1x_4x_7 - (u_1x_5 - u_1x_3)x_6 - u_1x_3x_4$

$R_8 = prem(R_9, f_8, x_8) = u_1^2x_4x_7 - (u_1^2x_5 - u_1^2x_3)x_6 - u_1^2x_3x_4$

$R_7 = prem(R_8, f_7, x_7) = -(u_1^2u_3x_5 + u_1^2u_2x_4 - u_1^2u_3x_3)x_6 + u_1^2u_2u_4x_4$

$R_6 = prem(R_7, f_6, x_6)$ (6 terms)[2]

$R_5 = prem(R_6, f_5, x_5)$ (6 terms)

$R_4 = prem(R_5, f_4, x_4)$ (11 terms)

$R_3 = prem(R_4, f_3, x_3)$ (11 terms)

$R_2 = prem(R_3, f_2, x_2)$ (8 terms)

$R_1 = prem(R_2, f_1, x_1) = 0.$

Since the final remainder R_1 is 0, by the remainder formula (2.5), $c = 0$ follows from $h_i = 0$ and subsidiary conditions $I_i \neq 0$, where the I_i are the leading coefficients of the f_i:

$$\begin{aligned}
I_1 &= 2u_1 \\
I_2 &= 2u_2 \\
I_3 &= 1 \\
I_4 &= u_3^2 - 2u_1u_3 + u_2^2 + u_1^2 \\
I_5 &= u_3 - u_1 \\
I_6 &= u_3^2 + u_2^2 \\
I_7 &= u_3 \\
I_8 &= I_9 = u_1.
\end{aligned}$$

[2] We omit the rest of the remainder polynomials, the largest of which has only 11 terms. Step 3 can be done in a fraction of one second on a Mac Plus. In view of the fact that the traditional proofs of Simson's theorem are non-trivial and subject to many gaps, the simple mechanical proof by the computer here is really a surprise.

Step 4. Analysis of subsidiary conditions $I_k \neq 0$. Here the non-zeroness of I_1, I_2, I_8, and I_9 means that A, B, and C are not collinear. $I_7 \neq 0$ and $I_5 \neq 0$ are not necessary by a more careful analysis based on the work [8] and [17]. The role of the conditions $I_4 \neq 0$ and $I_6 \neq 0$ is very subtle, and they are *necessary* when using Wu's method or the GB method. See Example (2.3) below; also see [17] for a further discussion. We would emphasize that all this analysis can be done mechanically and has been implemented [8] [17].

The reader might try to prove the theorem in Example (1.1) using the simple method described here.

Remark. This way of proving theorems is not the one adopted by most provers based on the logic approach. We start with a set of hypotheses which do not necessarily imply the conclusion and end up with the confirmation of the conclusion by adding some subsidiary conditions. Would it be possible that the original geometry statement is substantially weakened or changed? To answer this question we need to address the formulation problem – in what sense do we prove geometry theorems? Generally, for a geometry statement, the equality part of the hypotheses (i.e., $H = \{h_1 = 0, ..., h_r = 0\}$) is easy to identify. If some of them are missing and the statement is not confirmed (or disproved), it is the user's mistake. However, the inequation part of the hypotheses, which is usually connected with non-degeneracy, is not so simple to identify. For a given geometry statement, a non-degenerate condition that is obvious to one person might not be obvious to a second, and a third person might refuse to accept the condition as relevant or appropriate. The key issue here is how to understand and handle these non-degenerate conditions.

2.2. Geometry Statements of Constructive Type

This simple use of Wu's method is powerful enough to prove hundreds of nontrivial geometry statements. But its application is restrictive. First, the ndg conditions $I_i \neq 0$ depend on the choices of the coordinates and this may cause problems. See the 8_3 configuration problem in [14]. Secondly, $I_i \neq 0$ are in algebraic form and there is no general method to transform these conditions into geometric form.

The great success of Wu's method is closely connected to some special geometry statements: the class of constructive statements. Actually, the statements considered by Hilbert are constructive ones, but it has a rather restricted character: they are about those configurations formed by straight lines in a constructive way. In [8]and [17], we considered a class of constructive geometry statements (henceforth referred to as Class C) which are easy to prove and include most theorems in elementary geometry. Class C is actually the statements about the configurations which can be drawn by rulers and compasses. For instance, Example 2.1 can be described in the following constructive way: Points A, B, and C are arbitrarily chosen; point D is the intersection of the line passing through A and parallel to CB and the line passing through C and parallel to AB; E is the intersection of line BD and line AC. About 85 percent of the 512 theorems in [10] belong to this class.

The basic method introduced in Section 2.1 can be developed into a complete method for constructive statements. For a constructive statement, point coordinates can be chosen automatically and coordinate-independent ndg conditions in geometric form can be generated. Furthermore, these ndg conditions are sufficient, i.e., a geometry statement is true in a complex geometry iff it is true under these ndg conditions. For geometry statements of constructive type, since new points are introduced one by one, the new dependent variables are introduced at most two by two. Therefore their corresponding algebraic problems are easier than the algebraic problems corresponding to the general geometry problems.

A prover based on the simple version of the method for constructive statements in the Macintosh

Plus is available [15]. For example, the input of Simson's theorem to the prover is:

(2.8)
```
'((point-order A B C O D E F G)
  (collinear A B G)
  (collinear B C E)
  (collinear A C F)
  (perpendicular E D B C)
  (perpendicular F D A C)
  (perpendicular G D A B)
  (eqdistance A O B O)
  (eqdistance A O C O)
  (eqdistance A O D O)
  (collinear E F G))                    ;The last line is the conclusion
```

For Simson's theorem, the prover generates non-degenerate conditions (from the input (2.8)) that "the three vertices are not collinear" and "the three sides are not isotropic". An isotropic line is a line perpendicular to itself. The algebraic conditions for, say, BC and AC being non-isotropic, are $I_6 = (u_3 - u_1)^2 + u_2^2 \neq 0$ and $I_4 = u_3^2 + u_2^2 \neq 0$. Thus the exact statement of Simson's theorem can be:

$$\forall ux[(H \wedge D) \Rightarrow c = 0],$$

where H is the equality hypotheses and $D = (d_1 \neq 0 \wedge u_1^2 \neq 0 \wedge I_4 \neq 0 \wedge I_6 \neq 0)$.

More properties of constructive statements can be found in [17]. In [16], most of the results about constructive statements have been extended to Riemann geometry.

2.3. The Formulation Problem

Generally, for a geometry configuration defined by a set of equations

$$H = \{h_1(y_1, ..., y_m) = 0, ..., h_r(y_1, ..., y_m) = 0\},$$

we want to decide whether an assertion $c(y_1, ..., y_m) = 0$ on this configuration is valid. We define the notation

$$Zero(h_1, ..., h_r) = \{(y_1, ..., y_m) \in E^m \,|h_i(y_1, ..., y_m) = 0 \text{ for } i = 1, ..., r\},$$

where E is an extension of the base field $\mathbf{Q}$; usually, $E = \mathbf{C}$ (the field of complex numbers; later we will discuss the case when $E = \mathbf{R}$, the field of really numbers). As we know the formula $\forall y \in E[H \Rightarrow c = 0]$, or equivalently

$$Zero(H) \subset Zero(c)$$

is usually not valid because of missing non-degenerate conditions.[3]

It is now accepted by most researchers that there are two different but related formulations for dealing with non-degenerate conditions.

Formulation F1. Identify some variables among $y_1, ..., y_m$ as parameters, then decide whether the conclusion $c = 0$ follows from the hypothesis H generally (generically) with respect to those parameters. The zeros of the hypothesis polynomials can be decomposed into irreducible components (in an algebraically closed field) as follows:

$$Zero(H) = V_1^* \cup \cdots V_d^* \cup V_1^{\text{dege}} \cup \cdots \cup V_l^{\text{dege}},$$

[3] When the meaning is clear from context, we will use $f = 0$, f, and $\{f\}$ interchangeably.

where $V_1^*, ..., V_d^*$ are all those components on which the parameters are algebraically independent, corresponding to non-degenerate cases; the others correspond to degenerate cases. If the conclusion is valid in all non-degenerate cases, then we say the statement is *generally (generically) true*. If it is valid on none of the non-degenerate cases, then it is *generally false*. For most problems in practice, there is only one non-degenerate case (i.e., $d = 1$); then we always have a clear cut answer: it is either generally true or generally false.

If the conclusion is valid on some non-degenerate cases, but not all, then the statement needs further investigation. The problem usually becomes more interesting; for details see [10].

Formulation F2. Explicitly specify the non-degenerate condition $D = \{d_1 \neq 0, ..., d_q \neq 0\}$ as an integral part of the hypotheses. Then the aim is to decide whether the statement

$$\forall y[(H \wedge D) \Rightarrow c = 0] \tag{2.9}$$

is valid without adding any additional conditions.

F1 can help to find the missing non-degenerate conditions. Furthermore, it addresses the nature of the statement: if a statement is proved to be generally false, it cannot be a theorem no matter how many reasonable additional non-degenerate conditions are added. However, non-degenerate conditions are usually in algebraic form and the currently used methods based on F1 usually generate non-degenerate conditions more than needed. On the other hand, F2 is easier to understand. The geometry statement is exactly specified, and the user can select non-degenerate conditions he/she thinks suitable. However, if the statement is proved to be false, we don't know the nature of the statement: whether it is generally false or the proof failed due to missing non-degenerate conditions. In my experience non-degenerate conditions are often implicit in a statement in geometry textbooks and identifying them is sometimes very hard and subtle, even in Euclidean geometry; see [17] for details. Note that the equality part of the hypotheses (i.e., H) are *the same* for a given specification of a geometry statement regardless of how non-degenerate conditions are added or understood. For a detailed discussion of the formulation problem, see [21]; B. Kutzler, expresses a different point of view in [38] [39].

Now we come to the question why we chose E to be $\mathbf{C}$, the field of complex numbers, instead of $\mathbf{R}$, the field of real numbers, which is considered more realistic by some authors. First, we emphasize that if the CS or GB method confirms a geometry statement either according to F1 or F2, then it is valid in *all* fields, including $\mathbf{R}$. However, if the statement is disproved by one of the two methods, then it is not valid in $\mathbf{C}$, but may be valid in $\mathbf{R}$. In axiomatic geometry, there are several systems of axioms. There is a system of axioms for *unordered metric geometry* which involves the relations incidence, perpendicularity, segment congruence, and angle congruence; $\mathbf{C}^2$ (or $\mathbf{R}^2$) is a *model* for the theory of this metric geometry [52] [8]. If we want to decide whether a geometry statement (with universal quantifiers outside) is a *logical consequence* of the theory of metric geometry, then the CS and GB methods are complete, i.e., they are a *decision procedure* for such theory.

Example (2.3). (Simson's theorem continues). Suppose we assume the only non-degenerate condition is "A, B, and C are not collinear", i.e., $d_1 = u_1 u_2 \neq 0$ for Simson's theorem. Then we ask (according to F2) whether the statement is a logical consequence of the theory of metric geometry, or equivalently, in its algebraic form, whether

$$\forall ux \in \mathbf{C}[(H \wedge d_1 \neq 0) \Rightarrow c = 0]$$

is valid. Then the CS or GB method proves it is not the case. That means the statement, as it is, is *not a theorem* (logical consequence) in the theory of metric geometry. Or putting it another way, the statement cannot be proved *without using axioms of order*. However, if we add another non-degenerate condition that "the three sides of the triangle are non-isotropic", the statement is true as verified by both methods. In Euclidean geometry, isotropic lines do not exist, thus Simson's theorem is proved by the CS or GB method under the sole non-degenerate condition that A, B and C are not collinear.

2.4. Methods

From the above discussion, we have four approaches: CS(F1), CS(F2), GB(F1), and GB(F2). According to D. Kapur, one can use direct or refutational approaches. Thus altogether there are possibly $4 \times 2 = 8$ approaches. Here we cite the most recent presentations for those approaches: for CS(F1), see [10]; for CS(F2) see [36] [11] [33]; for GB(F1), see [20] [37] [10] [38]; for GB(F2), see [20] [32] [10] [31]. We now briefly present a representative of each of the four approaches.

CS(F1) [7] [10]. The method we just used for the two examples in Section 2.1 is actually based on Formulation F1. In both examples, the triangular forms ASC are irreducible and represent the only non-degenerate component, V_1^*, and $prem(c; ASC) = 0$ means that $c = 0$ is valid on V_1^*, i.e., the statements are generally true. For a complete description of CS(F1), see [10].

CS(F2) [36] [11]. Let S and G be two polynomial sets. Denote $Zero(S/G)$ the set difference $Zero(S) - \bigcup_{d \in G} Zero(d)$. Thus according to F2, our goal is to prove (2.9), i.e., to prove

$$Zero(H/D) \subset Zero(c),$$

where $H = \{h_1 = 0, ..., h_r = 0\}$ and $D = \{d_1 \neq 0, ..., d_q \neq 0\}$ are the equality part and the inequation part of the hypotheses, respectively.

Using Ritt–Wu's decomposition algorithm [11],

$$Zero(H/D) = \bigcup_{1 \le i \le k} Zero(PD(ASC_i)/D)$$

where the ASC_i are irreducible ascending chains; the notation $PD(ASC)$ is defined to be

$$\{g \mid prem(g, ASC) = 0\}.$$

By theorem (5.5) in [11], to decide whether the statement is true or not (in **C**), we only need to verify whether $prem(c; ASC_i) = 0$ for all i. If we chose $D = \{u_1 u_2 \neq 0\}$ for the parallelogram example and D to be the same as in Example (2.3) for Simson's theorem, then $k = 1$, and ASC_1 are just the triangular forms ASC in (2.4) and (2.7). Since $prem(c; ASC_1) = 0$ in both examples (Step 3: successive division), they have been confirmed under the non-degenerate conditions specified by us without adding any other conditions.

GB(F1) [20] [37]. The conclusion $c = 0$ follows from the hypotheses $h_1 = 0, ..., h_r = 0$ generically iff there is a non–zero polynomial U containing only the parameters such that $U \cdot c \in Radical(h_1, ..., h_r)$ (Theorem 7.1 in [21]). This is in turn equivalent to c being in the radical generated by $h_1, ..., h_r$ in the ring $\mathbf{Q}(u)[x]$, where the u are parameters, and the x are the dependent variables. As proved in [10], this is equivalent to a Gröbner basis of $h_1, ..., h_r, cz - 1$ in $\mathbf{Q}(u)[x]$ containing 1, where z is a new variable. This is the case for Simson's theorem as confirmed by our computer program, and the polynomial U was also found during computing the GB. Thus under $U \neq 0$, $(H \Rightarrow c = 0)$. It is generally hard to interpret the geometric meaning of $U \neq 0$ automatically. However, the following theorem solves this problem indirectly for statements in Class C.

Theorem (2.4). [8] [17]. For a constructive statement, if $(H \Rightarrow c = 0)$ is confirmed to be generally true, then $(H \Rightarrow c = 0)$ is valid under the *geometric* non-degenerate conditions generated by the algorithm in [17].

Based on this this theorem, Simson's theorem has been proved by GB(F1) under the non-degenerate conditions that the three vertices are not collinear and the three sides are non-isotropic.

GB(F2). [10] [32]. First we observe that in any field, $d \neq 0$ iff $\exists z(zd-1=0)$. Thus (2.10) is equivalent to

$$\forall y[\exists z_1 \cdots z_q(H \wedge d_1 z_1 - 1 = 0 \wedge \cdots \wedge d_q z_q - 1 = 0) \Rightarrow c = 0],$$

which is in turn equivalent (because c is free of z_i) to

$$\forall y z_1 \cdots z_q[(H \wedge d_1 z_1 - 1 = 0 \wedge \cdots \wedge d_q z_q - 1 = 0) \Rightarrow c = 0].$$

As proved in [10] [31], in algebraically closed fields, the above formula is equivalent to whether $H' = \{h_1, ..., h_r, d_1 z_1 - 1, ..., d_q z_q - 1, zc - 1\}$ generates the unit ideal, where the z are new variables. Thus the method in [31] and [10] is to compute a Gröbner basis of H' to see whether it contains 1. It is the case as confirmed by computers for our two examples. The use of new variables z_i and z was first introduced by Rabinowitsch in connection with a proof of Hilbert's Nullstellensatz. It has been used extensively in computer algebra, e.g., in [28], since 1983. In geometry theorem proving, it was first used by Chou and Schelter [20], and Kapur [31].

2.5. A Comparison of CS and GB

These two methods address the same geometry statements of equality type and are complete only for complex geometry. If a geometry statement is proved in complex geometry, it is also valid in Euclidean geometry (or in any other field). The converse is generally not true. In [10] such an example was given. A more natural example which is valid in **R**, but not in **C**, the field of complex numbers, is the 8_3 problem [38],[14]. A simple algebraic example is

$$\forall u_1 x_1[u_1^2 + x_1^2 = 0 \Rightarrow x_1 = 0].$$

Both CS and GB method can disprove this formula in **C**, but cannot confirm it in **R**. However, such kinds of formulas rarely happen in Euclidean geometry. If we change the above formula a "little" bit:

$$\forall u_1 x_1[u_1^2 + x_1^2 - 1 = 0 \Rightarrow x_1 = 0],$$

then for $u_1 \in (-1,1)$, x_1 always has solutions in **R**. Such kinds of formulas are called **R**–generic. If an **R**–genéric formula is not valid in **C**, it is also not valid in **R**. Most geometry statements are **R**–generic, and for such statements the CS method is complete for Euclidean geometry. For details, see [21] [10] [35]. This is *the real reason* that the CS method, which is complete only for complex geometry, can prove so many theorems in real geometry. One of my own contributions to the CS method has been to study reducible geometry problems extensively (i.e., there are more than one non-degenerate components; see Thèbault's conjecture in [10]). There has been no similar work done for the GB method.

The current implementations show the CS method is generally faster. For 413 of the 512 theorems in [10] proved by CS(F1), all non-degenerate conditions are in geometric form. About 90–95% were proved by CS(F2) and GB(F1). Thus nearly 500 theorems were proved in their full geometric form. The following table gives the time statistics of the four approaches using our implementations. A sign (*) means that some non-degenerate conditions are still in algebraic form. A sign (?) means that it went more than 4 hours without results.

No	Theorem	Sources	CS(F1)	CS(F2)	GB(F1)	GB(F2)
E1	Parallelogram	[20]	0.1	0.25	0.45	0.71
E2	Centroid	E1 [20]	0.35	0.35	0.83	0.38
E3	Altitudes	E133 [10]	0.83	0.6	2.5	75.7
E4	Ceva's Th	E334 [10]	0.81	0.7	1.1	48.3
E5	Simson's Th	E2 [20]	1.2	0.7	1.6	144.1
E6	Brahmagupta	[20]	1.1	5.2	2.4	1.48
E7	Pappus' Th	E6 [20]	1.3	2.7	6.4	44.6
E8	Pappus Dual	E17 [10]	1.7	1.1	15.8	293.9
E9	Butterfly	E5 [20]	1.5	6.2	24.5	?
E10	Simson Dual	E21 [10]	4.8	21.1	?	?
E11	Pascal's th	[11]	3.9	9.2	1,401	?
E12	Brianchon	E19 [10]	9.6	10.3	?	?
E13	Pappus Point	E7 [20]	5.6	10.7	54.6	?
E14	Isosceles	E8 [20]	2.6	3.8	1.5	3.3
E15	Gauss' Th	E9 [20]	1.5	0.20	1.3	0.35
E16	Gauss Point	E10 [20]	3.7	6.3	18.8	95.1
E17	Gauss Conic	E17 [10]	1,511	1,603	5,225	?
E18	Feuerbach	E204 [10]	2.8	28.3	28.2	197.9
E19	Miquel Point	E308 [10]	40.2	44.4	90.3	?
E20	Miquel Circle	E309 [10]	412.3	559.1	137.7	?
E21	Steiner's Th	E13 [10]	10.2	12.8	?	?
E22	Pas-Conic 1	E9 [10]	11.8	14.0	71.6	?
E23	Morley's Th	[11]	20.2 *	763.1	308.1 *	?
E24	V. Pratt's Th	E80 [10]	275.3	?	?	?
E25	Coxeter's Th	E40 [10]	218.2 *	?	?	?

Here we state E23 (Morley's trisector theorem): The points of intersection of the adjacent trisectors of the angles of any triangle are the vertices of an equilateral triangle (Fig. 5). This theorem has been proved under the non-degenerate conditions that the three vertices are not collinear and the three sides are non-isotropic using CS(F2) [11]. CS(F1) and GB(F1) confirmed it to be generally true, but with non-degenerate conditions in algebraic form [10].

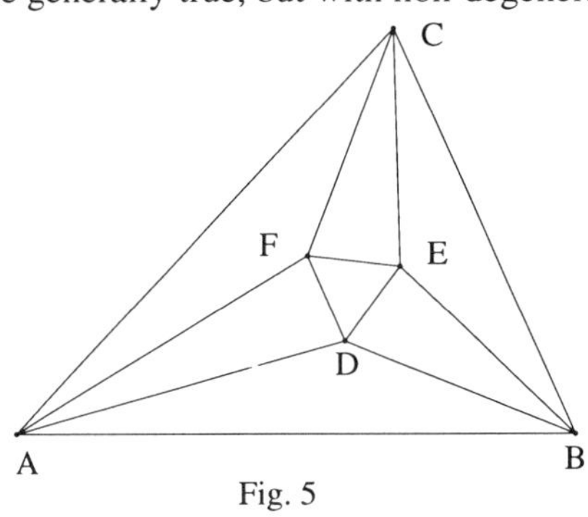

Fig. 5

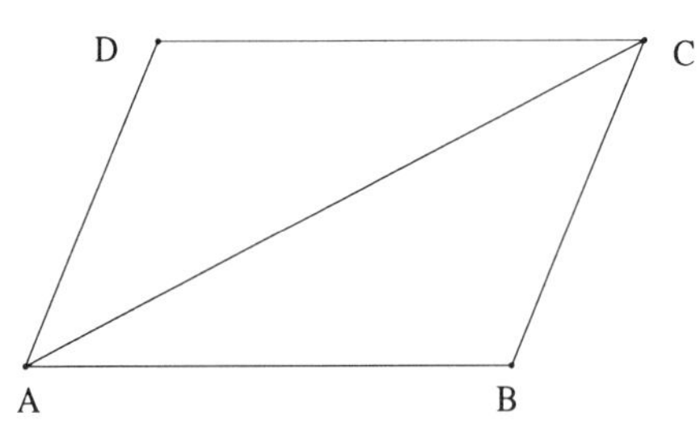

Fig. 6

2.6. Proving Geometry Theorems by Examples

Based on the CS method, J. Hong showed that to prove a geometry statement, we only need to check a single numerical example [30]. To understand the method, let us mention the simple fact: a polynomial $p(x_1) \in Q[x_1]$ is identically zero if $p(x_0) = 0$ for a sufficiently large rational number x_0. Hong's work is actually a generalization of the above result.

In [63] [62], a method of proving geometry theorems by checking several numerical examples instead of one is presented. This method is similar to Hong's method and is based on a generalization of the following fact: a polynomial $p(x_1)$ of degree d is identically zero if it has more than d distinct roots.

In both of the above methods, we need to do approximate calculations and at last we need to check whether an approximate number is small enough to be zero which is a difficult problem. However, in the case of linear geometry statements, the approximation problem can be avoided by using arbitraryly large rational number calculation which is widely available in symbolic computer softwares. A prover for linear statements has been developed and has been used to prove many nontrivial examples [62].

2.7. Proving Theorems with Inequalities

In elementary geometry, there are also many theorems involving inequalities (e.g., Example (1.2)). Proving theorems with inequalities is within the scope of Collins' method, but beyond the scope of the CS and GB methods. Theoretically, Collins' method can prove (or disprove) any elementary sentences in the Tarski geometry. A theorem involving inequalities generally also involves equalities. Since the CS and GB methods work so well for equality problems, we might expect a combination of the CS (or GB) method with Collins' method could solve problems not in the scope of the CS method, but which cannot be solved by Collins' method alone within the available time and space. The work in [3] is in this direction, and a number of hard problems were solved with some human interaction. Here we use the following simple example to illustrate the basic idea.

Example (2.5). Let $ABCD$ be a parallelogram. Show that points B and D are on either side of diagonal AC (Fig. 6).

This "trivial" fact is repeatedly used in traditional proofs of the properties of a parallelogram, e.g., in proving the two diagonals bisect each other (Example 2.1). However, it seems nontrivial to find a rigorous traditional proof of this fact. (Try it!)

Let $A = (0, 0)$, $B = (u_1, 0)$, $C = (u_2, u_3)$, and $D = (x_2, x_1)$. Then we have two equations for the hypotheses

$$h_1 = u_1x_1 - u_1u_3 = 0 \qquad AB \text{ is parallel to } CD$$
$$h_2 = u_3x_2 - (u_2 - u_1)x_1 = 0 \qquad AD \text{ is parallel to } BC.$$

The conclusion that B and D are on either side of AC is $g < 0$, where $g = (u_3u_1 - u_2 \cdot 0)(u_3x_2 - u_2x_1) = u_1u_3^2x_2 - u_1u_2u_3x_1$. We want to decide whether the following statement is valid under certain nondegeneracy conditions:

$$\forall u_1u_2u_3x_1x_2[(h_1 = 0 \wedge h_2 = 0) \Rightarrow g < 0].$$

Here u_1, u_2, u_3 are selected to be parameters, and x_1 and x_2 are selected to be dependent variables. Reducing g to canonical form modulo the ideal (in $\mathbf{Q}(u)[x]$) generated by h_1 and h_2 (this is called the "canonical simplification" of g with respect to this ideal by [5]), we obtain $g = -u_1^2u_3^2$. Alternatively, we can solve $x_1 = u_1$, $x_2 = u_2 - u_1$ and substitute the solution into g to get $g = -u_1^2u_3^2$. This canonical form of g modulo the ideal is only valid under the conditions $u_1 \neq 0$ and $u_3 \neq 0$; in other words, u_1 and u_3 occur in the denominators of the elements c_1 and c_2 of $Q(u)[x]$ such that $g = c_1h_1 + c_2h_2$. Thus we have $g < 0$, under the condition that $u_1u_3 \neq 0$. Note that $u_1u_3 \neq 0$ is indeed connected with nondegeneracy, i.e. to insure that points A, B and C are not collinear.

This technique can also solve Example (1.2) easily. A more general scheme has been proposed, and it has been used to solve the 8_3 problem automatically [14]. Recently, N. McPhee gives an automatical solution to the Steiner-Lehmus theorem and the Pompieu's theorem using a similar approach [43].

3. Proving Theorems in Differential Geometry

Here we are dealing with differential polynomial rings over a differential field (usually it is $\mathbf{Q}(t)$) in which there is a third operation, "′", compatible with the two operations of a ring, "+" and "*":

$$(a+b)' = a' + b'$$
$$(ab)' = a'b + ab'.$$

The complete CS method for differential polynomials has been developed. There is an algorithm for deciding whether a differential polynomial is in the *radical* ideal generated by a finite set of differential polynomials [45], and the pseudo division, triangular algorithm, and the variety decomposition algorithm can be extended to the differential polynomial case with minor modifications [45] [53] [18]. In fact, the CS method for ordinary polynomials is a special case of differential polynomials.

3.1. Space Curves and Mechanics

The geometry statements addressed are still of *equality type*. In the local theory of space curves, one uses parametric representation of a curve: $C = (x(t), y(t), z(t))$. The practical problems we encountered in the curve theory and mechanics are of the similar nature. Thus we use the following elegant example first worked by Wu [57] to illustrate the type of problems we address:

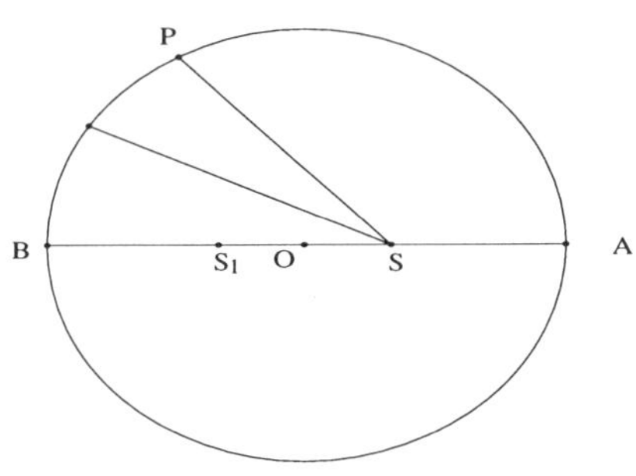

Fig. 7

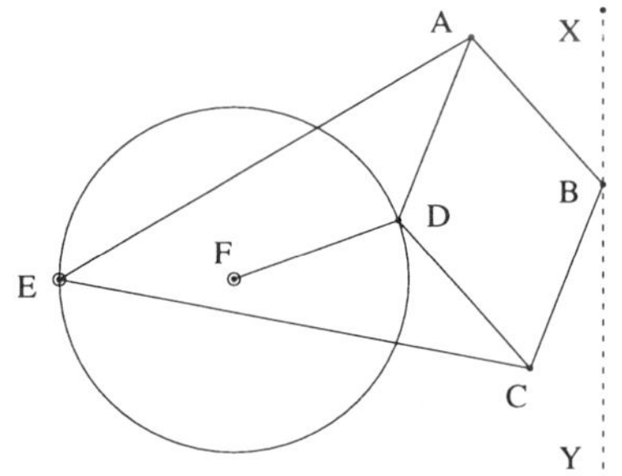

Fig. 8

Example (3.1). The Kepler–Newton problem.

Kepler's first two laws are:

(K1) The planets move in elliptic orbits with the sun as a focus (Fig. 7).

(K2) The vector from the sun to the planet sweeps equal areas in equal times.

Newton's law of gravitation (special form):

(N1) The acceleration of a planet is inversely proportional to the square of the distance from the sun to the planet.

Now we want to prove that (K1) and (K2) imply (N1).

Choose the sun as the origin of the coordinate system, and let $(x(t), y(t))$ be the position of the planet; $(-c, 0)$ be the center of the ellipse; h be the area velocity of the planet. Then the equation part of the hypotheses is:

$K1_1$: $$\frac{(x+c)^2}{a^2} + \frac{y^2}{b^2} - 1 = 0$$

$K1_2$: $$a^2 - (b^2 + c^2) = 0$$

$K2$: $$x'y - xy' - h = 0, \qquad \text{with} \quad h' = 0.$$

f_1 : $$r^2 - (x^2 + y^2) = 0$$

f_2 : $$A^2 - (x''^2 + y''^2) = 0$$

The conclusion N1 is: $[Ar^2]' = 0.$

Then a tentative algebraic form for this problem would be:

$$\forall v[(K1_1 \wedge K1_2 \wedge K2 \wedge f_1 \wedge f_2) \Rightarrow \mathrm{N1}].$$

As usual, x' denotes the derivative of x with respect to t. Here, a, b, c, and h are constants (independent of t), i.e., their derivatives are 0. Again, the above formula is valid under some non-degenerate conditions. With the simple use of the CS method described in Section 2.1, Wu proved a variation of the above specification [57] under some non-degenerate conditions. However, in Wu's work the formulation of the problem is not clear.

As in elementary geometry, there are also two Formulations F1 and F2 for proving theorem in differential geometry. While these two Formulations and the CS methods for them are similar to the case of elementary geometry, Formulation F1 is much more complicated than a simple rewording. Chou and Gao have developed methods for both F1 and F2 [19], and the above example has been proved using both Formulations. Nearly 100 theorems in space curves and 10 statements in mechanics have been proved [18].

3.2. Space Surfaces

This involves partial differential polynomials (pdp). In this area, Wu has developed tools [58] [54] [53] [59] based on the work of Ritt [45], J. M. Thomas [48], and E. Cartan [6], where again the method is only for geometry statements of equality type. This is similar to, but much more complicated than the ordinary differential polynomial case. Some experiment results are give in [41].

The method currently used is the simple one described in Section 2.1, i.e., first triangulize a pdp set to obtain a pdp ascending chain ASC, then do similar successive pseudo divisions of the conclusion pdp with respect to that ASC to see whether the final remainder is zero. In elementary geometry such simple use of the CS method often proves a geometry statement to be generally true because most statements are of constructive types. However, in the pdp case, the situation is unclear. There is no work to formulate "generally true" precisely as in Formulation F1 for elementary geometry. Formulation F2 is straight forward and the CS method for F2 can presumably be given, though no experimental work has been done yet.

4. Geometry Formula Derivation

4.1. Elementary Geometry

There are two kinds of problems in elementary geometry other than theorem proving. One is finding locus equations, the other is deriving geometry formulas. The first problem was addressed in [7] [8] [26]; the second in [55] [9]. Actually, these two kinds of problems can be formulated in a unified way as it has been done in [12]. We use Heron's Formula to illustrate this type of problem.

Example (4.1). Find the formula for the area of a triangle ABC in terms of its three sides.

Let a, b, and c be the three sides, $B = (0,0)$, $C = (a,0)$, and $A = (x_1, x_2)$. Then the geometry conditions can be expressed by the following set of polynomial equations:

$$\begin{array}{ll} h_1 = x_2^2 + x_1^2 - 2ax_1 - b^2 + a^2 = 0 & b = AC \\ h_2 = x_2^2 + x_1^2 - c^2 = 0 & c = AB \\ h_3 = ax_2 - 2k = 0 & k = \text{the area of } ABC. \end{array}$$

The aim is to find a polynomial equation involving only a, b, c, and k which is a consequence of the above equations and some non-degenerate conditions.

In general, for a geometric configuration given by a set of polynomial equations $h_1(u_1, ..., u_q, x_1, ..., x_p) = 0, ..., h_r(u_1, ..., u_q, x_1, ..., x_p) = 0$ (possibly with a set of inequations $\{D = d_1(u,x) \neq 0, ..., d_s(u,x) \neq 0\}$), we want to find a relation (formula) between arbitrarily chosen variables $u_1, ..., u_q$ (parameters) and a dependent variable, say, x_1. The work in [55] [8] [9] aimed to address this problem. In [12], two methods based on the GB method and one method based on the CS method are given, which, in our opinion, are more satisfactory. Heron's formula can be easily derived by any of the above methods:

$$16k^2 + c^4 - (2b^2 + 2a^2)c^2 + b^4 - 2a^2b^2 + a^4 = 0.$$

The two methods based on the GB method are conceptually simpler. They are equivalent to finding a polynomial in only the u_j and x_1 with the least positive degree of x_1 contained in the ideals generated by $h_1, ..., h_s$ in $\mathbf{Q}[u, x]$ and $\mathbf{Q}(u)[x]$. Here is a more interesting example.

Example (4.2). (Peaucellier's Linkage). Links AD, AB, DC and BC have equal length, as do links EA and EC. The length of FD equals the distance from E to F. The locations of joints E and F are fixed points on the plane, but the linkage is allowed to rotate about these points. As it does, what is the locus of the joint B? (Fig. 8)

Let $F = (0,0)$, $E = (r,0)$, $C = (x_2, y_2)$, $D = (x_1, y_1)$, and $B = (x, y)$, n and m be the lengths of the projections of AD on the direction of line EDB (note that points E, D, and B are always collinear) and on the direction perpendicular to it. Then the geometry conditions can be expressed by the following set of polynomial equations H

$$\begin{array}{ll} h_1 = y_1^2 + x_1^2 - r^2 = 0 & r = FD \\ h_2 = y_2^2 - 2y_1y_2 + x_2^2 - 2x_1x_2 + y_1^2 + x_1^2 - n^2 - m^2 = 0 & CD = n^2 + m^2 \\ h_3 = y_2^2 - 2yy_2 + x_2^2 - 2xx_2 + x^2 + y^2 - n^2 - m^2 = 0 & CB = n^2 + m^2 \\ h_4 = y_2^2 + x_2^2 - 2rx_2 - n^2 - 4rn - m^2 - 3r^2 = 0 & EC = (n + 2r)^2 + m^2 \\ h_5 = (x - r)y_1 - yx_1 + ry = 0 & E \text{ is on } DB, \end{array}$$

together with the following set of polynomial inequations D:

$$d_1 = x_1 - x \neq 0 \qquad B \neq D.$$

Selecting m, n, r, and y to be the parameters of the problem, we want to find the relation among m, n, r, y and x. Using Method 4.6 in [12] (in the ordering $x < x_1 < y_1 < x_2 < y_2$), we have found $Zero(H/D)$ has only one non-degenerate component with the corresponding ascending chain $ASC_1^* =$

$$\begin{array}{l} x + 2n + r \\ (x^2 - 2rx + y^2 + r^2)x_1 + rx^2 - 2r^2x - ry^2 + r^3 \end{array}$$

$$\begin{aligned}
&(x-r)y_1 - yx_1 + ry \\
&(4x^2 - 8rx + 4y^2 + 4r^2)x_2^2 + (-4x^3 + 4rx^2 + (-4y^2 - 16rn - 12r^2)x - 4ry^2 + 16r^2n + 12r^3)x_2 \\
&\qquad + x^4 + (2y^2 + 8rn + 6r^2)x^2 + y^4 + (-4n^2 - 8rn - 4m^2 - 6r^2)y^2 + 16r^2n^2 + 24r^3n + 9r^4 \\
&2yy_2 + (2x - 2r)x_2 - x^2 - y^2 - 4rn - 3r^2.
\end{aligned}$$

The first polynomial is the relation $x + 2n + r = 0$, which tells us that the locus is a line parallel to the y-axis.

This problem can also be formulated as one of finding a quantifier free formula $f(u, x_1)$ such that $f(u, x_1) \iff \exists x_2 \cdots x_p[h_1(u, x) \wedge \cdots \wedge h_r(u, x) \wedge d_1(u, x) \neq 0 \wedge \cdots \wedge d_s(u, x) \neq 0]$. Formulated in this way, it is actually to calculate the projection of an algebraic set in the affine space E^{q+p} into the subspace E^{q+1}. If E is algebraically closed, there are methods for computing such projections. The method in [60] is based on the CS method. It would be interesting to explore this new formulation. If E is a real closed field, Collins' method gives a solution to the above problem; so it would also be interesting to examine the connection between the real closed case and algebraically closed case. For example, we expect that Collins' method produces the following form for $f(u, x_1)$ for Peaucellier's Linkage:

$$(x + 2n + r = 0) \wedge (-d \leq y \leq d) \wedge (\text{other nondegenerate conditions})$$

where d is from the 4th (quadratic) equation of the above ASC_1^*.

4.2. Differential Geometry and Mechanics

Formula derivation in differential geometry was initiated by Wu in connection with finding possibly unknown properties on Bertrand curves [61]. The approach used by Wu was to look at the partial differential polynomials produced during generation of a CS. This involved human assistance.

A more automatic method has been proposed [19] based on our methods in [12]. Using this method a complete list of the properties of Bertrand curves in metric and affine geometries has been obtained. About 10 examples in mechanics, including the Newton–Kepler problem, were successfully completed on computers; for details, see [19].

Acknowledgment. The first author wishes to thank N. McPhee for his assistance, and Professor G. Collins, H. Hong, and J. Johnson for the availability of Hong's partial CAD program.

5. Appendix

5.1. Pseudo Division

Let

$$\begin{aligned}
g &= a_n y^n + \cdots + a_1 y + a_0 \qquad && (a_n \neq 0) \\
f &= b_k y^k + \cdots + b_1 y + b_0 \qquad && (b_k \neq 0 \wedge k > 0)
\end{aligned}$$

The pseudo division algorithm is as follows:

```
bk := lc(f, y);     k := ld(f, y);
r := g;             n := ld(r, y);

while n ≥ k do
    begin
      an := lc(r, y);
      r := bk r − an y^(n−k) f;
      n := ld(r, y)
    end.
```

Since n strictly decreases after each iteration, the above while loop will terminate in at most $s \leq n - k + 1$ steps. Upon termination we have the pseudo remainder $r = prem(g, f, y)$. It is easy to prove the remainder formula (2.3).

Added in proof.

Recently, an entirely new method (the area method) for proving geometry theorems was introduced [64][65][66][67]. The area method is based on a few geometry invariants, such as area, ratio of segments, Pythagorean difference, etc. The main advantage of the new method is that the proofs generated by the method generally are short and human-readable. The computer program based on the area method has produced proofs of more than 500 geometry theorems, some of which are even shorter than those given by geometry experts. For more details, see the newly published book [68].

References

[1] D. Arnon and M. Mignotte, "On Mechanical Quantifier Elimination for Elementary Algebra and Geometry", *J. Symbolic Computation*, **5** (1988), 237-259.

[2] D. Arnon, "Geometric Reasoning with Logic and Algebra", *Artificial Intelligence Journal*, V. **37** (1988), pp. 37-60.

[3] D. Arnon, S.C. Chou and X.S. Gao, "On the Mechanical Proof of Geometry Theorems Involving Inequalities", "*Issues in Robotics and Nonlinear Geometry*",*Advances in Computing Research*, vol. 6, (ed. C.M. Hoffmann), p. 139-181, 1992, JAI Press INc, Greenwich, USA. .

[4] B. Buchberger, "An Algorithm Criterion for the Solvability of a System of Algebraic Equations" (German), *Aequationes Math.*, **4** (1970), 374-383.

[5] B. Buchberger, "Gröbner Bases: An Algorithmic Method in Polynomial Ideal Theory", Chapter 6 in *Recent Trends in Multidimensional Systems Theory*, N.K. Bose (ed.) D. Reidel Publ. Comp. 1985.

[6] E. Cartan, "Les Systemes differentiels exterieurs et leurs applications geometriques", Paris, 1945.

[7] S.C. Chou, "Proving Elementary Geometry Theorems Using Wu's Algorithm", in *Automated Theorem Proving: After 25 years*, Ed. By W.W. Bledsoe and D. Loveland, AMS Contemporary Mathematics Series **29** (1984), 243-286.

[8] S.C. Chou, "Proving and Discovering Theorems in Elementary Geometries Using Wu's Method", PhD Thesis, Department of Mathematics, University of Texas, Austin (1985).

[9] S.C. Chou, "A Method for Mechanical Derivation of Formulas in Elementary Geometry", *Journal of Automated Reasoning*, **3**(1987), 291–299.

[10] S.C. Chou, *Mechanical Geometry Theorem Proving,* D. Reidel Publishing Company, Dordrecht, Netherlands, 1988.

[11] S. C. Chou and X. S. Gao, "Ritt-Wu's Decomposition Algorithm and Geometry Theorem Proving", in *Proceedings of CADE-10, Lecture Notes in Computer Science*, Vol. 449, 207–220, 1990.

[12] S. C. Chou and X. S. Gao, "Methods for Mechanical Geometry Formula Deriving", in *1990 Proceedings of International Symposium on Symbolic and Algebraic Computation* ACM Press, 265–270.

[13] S.C. Chou and X.S. Gao, "A Collection of 120 Computer Solved Geometry Problems in Mechanical Formula Derivation", TR-89-22, Computer Sciences Department, The University of Texas at Austin, August 1989.

[14] N. McPhee, S.C. Chou, and X.S. Gao, "A Combination of Ritt–Wu's Method and Collins'

Method", *Proc. of CADE-12*, p.401-415, France, 1994.

[15] S.C. Chou, "A Geometry Theorem Prover for Macintoshes", in *Proceedings of CADE–11, Lecture Notes in Computer Science*, Vol. 607, 686–689, 1992.

[16] S.C. Chou and X.S. Gao, "Mechanical Theorem Proving in Riemann Goemetry", TR-90-3, CS Department, The University of Texas at Austin, 1990, to appear in *Proc. of IWAR, 1992.*

[17] S.C. Chou and X. S. Gao, " Proving Geometry Statements of Constructive type", in *Proceedings of CADE–11, Lecture Notes in Computer Science*, Vol. 607, 20–34, 1992.

[18] S. C. Chou and X. S. Gao, "Automated Reasoning in Differential Geometry and Mechanics Part I and Part II", *Journal of Automated Reasoning*, 10:161-189, 1993.

[19] S. C. Chou and X. S. Gao, "Automated Reasoning in Differential Geometry and Mechanics Part III", *IFIP Transaction on Automated Reasoning*, p.1-12, North-Hollan.

[20] S.C. Chou and W.F. Schelter, "Proving Geometry Theorems with Rewrite Rules", *Journal of Automated Reasoning*, **2(4)** (1986), 253–273.

[21] S.C. Chou and G.J. Yang, "On the Algebraic Formulation of Certain Geometry Statements and Mechanical Geometry Theorem Proving", *Algorithmica*, Vol. **4**, 1989, 237–262.

[22] G.E. Collins, "Quantifier Elimination for Real Closed Fields by Cylindrical Algebraic Decomposition", *Lecture Notes In Computer Science*, **33** (1975), Springer-Verlag, Berlin, 134-183.

[23] G.E. Collins and H. Hong, "Partial CAD Construction in Quantifier Elimination", 1989, to appear in *Journal of Symbolic Computation.*

[24] P. J. Davis and E. Cerutti, "FORMAC meets Pappus", *The American Mathematical Monthly*, **76** (1969), 895–905.

[25] Xiaoshan Gao, "Transcendental functions and Mechanical Theorem Proving in Elementary Geometries", *Journal of Automated Reasoning*, 6:403-417, 1990.

[26] Xiaoshan Gao and Dongming Wang, "Geometry Theorems Proved Mechanically Using Wu's Method", Part on Elementary Geometries, MM preprint No2, 1987.

[27] H. Gelernter, J.R. Hanson, and D.W. Loveland, "Empirical explorations of the geometry-theorem proving machine", *Proc. West. Joint Computer Conf.*, 143-147, 1960.

[28] P. Gianni, B. Trager and G. Zacharias, "Gröbner Bases and Primary Decomposition of Polynomial Ideals", *Journal of Symbolic Computation*, Vol. 6, 1988, 149–169.

[29] D. Hilbert, *Foundations of Geometry*, Open Court Publishing Company, La Salla, Illinois, 1971.

[30] J.W. Hong, *"Can Geometry be Proved by an Example?" Scientia Sinica*, 29(1986), 824–834.

[31] D. Kapur, "Geometry Theorem Proving Using Hilbert's Nullstellensatz", in Proceedings of the 1986 Symposium on Symbolic and Algebraic Computation, 202-208.

[32] D. Kapur, "A Refutational Approach to Geometry Theorem Proving", *Artificial Intelligence*, Vol. 37, (1988) pp61–93.

[33] D. Kapur and Hoi K. Wan, "Refutational Proofs of Geometry Theorems via Characteristic Set Computation", in *Proceedings of International Symposium on Symbolic and Algebraic Computation*, ACM Press, 1990, 277–284.

[34] H.P. Ko and Moayyed A. Hussain, "ALGE-Prover – an Algebraic Geometry Theorem Proving Software", Technical Report, 85CRD139, General Electric Company, 1985.

[35] H.P. Ko and S.C. Chou, " A Decision Method for Certain Algebraic Geometry Problems", *Rocky Mountain Journal of Mathematics*, **19(3)**, 1989, 709–724.

[36] H.P. Ko, "Geometry Theorem Proving by Decomposition of Quasi-Algebraic Sets: An Application of the Ritt-Wu Principle", *Artificial Intelligence*, Vol. 37, pp95-122 (1988).

[37] B. Kutzler and S. Stifter, "Automated Geometry Theorem Proving Using Buchberger's Algorithm", in Proceedings of the 1986 Symposium on Symbolic and Algebraic Compu-

tation, 209-214.

[38] B. Kutzler, "Careful Algebraic Translations of Geometry Theorems", in *Proceedings of the 1989 Symposium on Symbolic and Algebraic Computation*, ACM, New York, 254–263.

[39] B. Kutzler, "Algebraic Approaches to Automated Geometry Theorem Proving", PhD Thesis, Johannes Kepler University, Linz, Austria, 1988.

[40] B. Kutzler and S. Stifter, "Collection of Computerized Proofs of Geometry Theorems", *RISC*, Tech. Rep. 86–12, Univ. Linz, Austria.

[41] Li Ziming,, "Mechanical Theorem-Proving of the Local Theory of the Surfaces", p.102-p120, MM-Preprints, NO. 6, MMRC.

[42] J.D. McCharen, R. A. Overbeek, L. A. Wos, "Problems and Experiments for and with Automated Theorem-Proving Programs", *IEEE Trans. on Computers* **C-25** (1976), 773-782.

[43] N. McPhee, "Mechanically Proving Geometry Theorems Using Wu's Method and Collins' Method", Phd Thesis, The University of Texas at Austin, 1992.

[44] J. F. Ritt, *Differential Equation from Algebraic Standpoint,* AMS Colloquium Publications Volume 14, New York, 1938.

[45] J. F. Ritt, *Differential Algebra,* AMS Colloquium Publications, New York, 1950.

[46] A. Seidenberg, "A New Decision Method for Elementary Algebra", *Annals of Math.*, **60** (1954), 365–371.

[47] A. Tarski, *A Decision Method for Elementary Algebra and Geometry,* Report R-109, second revised ed. Santa Monica, CA: The Rand Corporation, 1951.

[48] J. M. Thomas, "Differential Systems" Amer. Math. Soc. 1937.

[49] Wu Wen-tsün, "On the Decision Problem and the Mechanization of Theorem Proving in Elementary Geometry", *Scientia Sinica*, **21** (1978), 157-179.

[50] Wu Wen-tsün, "Basic Principles of Mechanical Theorem Proving in Geometries", *J. of Sys. Sci. and Math. Sci.* **4(3)**, 1984, 207-235, republished in *Journal of Automated Reasoning*, **2(4)** (1986), 221-252.

[51] Wu Wen-tsün, "Some Recent Advances in Mechanical Theorem-Proving of Geometries", in *Automated Theorem Proving: After 25 years*, American Mathematical Society, Contemporary Mathematics **29** (1984), 235-242.

[52] Wu Wen-tsün, *Basic Principles of Mechanical Theorem Proving in Geometries,* (in Chinese) Peking 1984.

[53] Wu Wen-tsün, "A constructive theory of differential algebraic geometry", Lecture Notes in Maht., No. 1255 (1987), Springer–Velag, 1987.

[54] Wu Wen-tsün, "Mechanical Theorem Proving in Elementary Geometry and Differential Geometry", *Proc. 1980 Beijing Symposium on Differential Geometry and Differential Equations* **2** (1982), 125-138, Science Press.

[55] Wu Wen-tsün, "A Mechanization Method of Geometry and its Applications – I: Distances, areas, and volumes", *J. of Sys. Sci. and Math. Sci.* Vol 6, 1986, 204-216.

[56] Wu Wen-tsün, "A Mechanization Method of Geometry and its Applications – II: Curve pairs of Betrand type", *Kexue Tongbao*, **32(9)** (1987), 535–538.

[57] Wu Wen-tsün, "Mechanical Derivation of Newton's Gravitational Laws from Kepler's Laws", *Research Preprints, Mathematics-Mechanization* No.1, pp53–61, 1987.

[58] Wu Wen-tsün, "Mechanical Theorem Proving in Elementary Differential Geometry", *Scientia Sinica*, Mathematics Supplement (I), 1979, 94–102 (in Chinese).

[59] Wu Wen-tsün, "On the Foundation of Algebraic Differential Geometry", *Research Preprints, Mathematics-Mechanization* No.3, pp1-26, 1989.

[60] Wu Wen-tsün, "On a Projection Theorem of Quasi–Varieties in Elimination Theory", *Research Preprints, Mathematics-Mechanization* No.4, pp40–53, 1989.

[61] Wu Wen-tsün *"A mechanization method of geometry and its applications II. Curve pairs of Bertrand type" Chinese Science Bulletin*, 17(1986), 1281–1284.

[62] L. Yang, J.Z. Zhang, & X.R. Hou, *"A Criterion of Dependency Between Algebraic Equations and Its Applications" Proc. of the 1992 international Workshop on Mechanization of Mathematics*, p.110-134, Inter. Academic Publishers, Beijing.

[63] J. Z. Zhang, L. Yang, and Mike Deng, "The Parallel Numerical Method of Mechanical Theorem Proving", *Theoretical Computer Science* 74, 253-271, 1990.

[64] S.C. Chou, X.S. Gao, and J.Z. Zhang, "Automated Production of Traditional Proofs for Constructive Geometry Theorems", *Proc. of Eighth IEEE Symposium on Logic in Computer Science*, p.48–56, IEEE Computer Society Press, 1993.

[65] S.C. Chou, X.S. Gao, and J.Z. Zhang, "Mechanical Geometry Theorem Proving by Vector Calculation", *Proc. of ISSAC-93, ACM Press*, p.284–291, 1993.

[66] J.Z. Zhang, S.C. Chou, and X.S. Gao, "Automated Production of Traditional Proofs for Theorems in Euclidean Geometry", *Annals of Mathematics and AI*, 13, (1995), p.109-137.

[67] S.C. Chou, X.S. Gao, and J.Z. Zhang, "Automated Production of Traditional Proofs in Solid Geometry", *Journal of Automated Reasoning*, 14, p.257-291, 1995.

[68] S.C. Chou, X.S. Gao, and J.Z. Zhang, *Machine Proofs in Geometry – Automated Production of Readable Proofs for Geometry Theorems,* World Scientific, Singapore, 1994.

Synthetic vs Analytic Geometry for Computers

Walter Whiteley§
Department of Mathematics and Statistics
York University
4700 Keele St, North York, Ontario, M3J-1P3, Canada
Whiteley@mathstat.YorkU.ca

Abstract

Computer geometry usually means analytic geometry. Analytic geometry usually means Cartesian coordinates and Euclidean geometry. We consider alternatives: synthetic geometries, such as projective geometry, and coordinate-free analytic geometries. Using classical invariants and Cayley algebra (extended exterior algebra), we describe translations from coordinate analytic geometry to coordinate-free 'invarant' analytic geometry and the unsolved problem of translating back to synthetic geometry. The goal is to include more appropriate geometry in computer-aided geometry - and produce 'better' proofs from Automated Theorem Provers.

1. Introduction

To teach geometry to a computer, we must know which 'geometry' we face. If there is a property to be recognized by a machine, we or the machine must find a representation for the property at a level of geometry which makes it simple enough to find [Whiteley 1995]. I present some observations about the levels of geometries, about their 'equivalent' representations (synthetic vs analytic), and about the form of computer proofs for geometry.

(A) Geometry comes in many layers - a hierarchy of levels from topology to Euclidean geometry. For human learning, for the expression of properties and for the selection of language, these layers of geometry are very different (Section 3). It is important to decide which geometry we want the computer to do! For computer learning, this is particularly important. Too low a level of geometry means too many parameters (irrelevant or redundant data), too many properties are available as hypotheses,

§ This work was supported by grants from FCAR (Québec) and NSERC (Canada).

and increased difficulty in learning. Too high a level of geometry means too few parameters, too few properties are available and learning has built in inaccuracy.

(B) Within each layer of the hierarchy of geometries, there are basic choices of how to express properties and proofs. As a basic illustration, we consider the extremes of analytic geometry and synthetic geometry for discrete configurations in projective space. For these two representations we can ask: (i) are they logically equivalent? (Section 4); (ii) what is the complexity of translating from one representation to the other? Is it linear? polynomial? exponential? non-constructive?(Section 5).

(C) For analytic geometries, proofs are 'best' expressed with identities - and all 'natural deductions' provide these identities if proper record keeping is used (Section 6). The central role of identities is expressed in Hilbert's Nullstellensatz - and its modern relative the Realstellensatz. This central role should be reflected in the computer presentation of geometric results in analytic geometry and algebraic versions of synthetic geometry.

(D) Some computer generated 'proofs' for geometric theorems are 'better' than others. As a geometer observing work on mechanically generated proofs, I encounter a number of different algorithms which generate proofs in elementry geometry (Euclidean, affine or projective). I propose some criteria for evaluating the geometric and logical qualities of the output of such proofs (Section 7).

These observations draw on twenty years of research on the foundations of analytic projective geometry (invariant theory), matroid theory and discrete geometry applied to rigidity, approximation theory, polyhedral pictures, etc. [Whiteley 1973, 1983, 1986, 1993]. In particular, I draw on many conversations within the Structural Topology Research Group at the Université de Montréal, and within an informal project to develop 'synthetic computational geometry' [Havel et al 1989, Sturmfels & Whiteley 1991, White 1991, Whiteley 1991]

In this interdisciplinary context, I risk some provocative conclusions based on scattered technical results in mathematics, personal observations and readings. Hopefully my polemic will stimulate debate.

2. The patterns of Analytic and Synthetic Geometry.

What do we mean by 'synthetic' and 'analytic' geometries? The difference between synthetic and analytic geometry resides in the language, in the models and in the constructions.

'Analytic geometry' is a 'geometry' whose language is based on the algebraic constructions - vector spaces over fields, addition, multiplication etc.. With this language go algebraic models and morphisms which preserve this algebra - appropriate linear transformations in most cases. The specific transformations depend on the 'geometry' (see the next section).

'Synthetic geometry' is a geometry whose language is based geometric constructions - intersecting lines, points on planes, equal distances etc.. With the geometric objects go morphisms which preserve these constructions - collineations, congruences .... The specific constructions depend on the 'geometry' chosen.

To match the fundamental geometry of vision, as well as classical invariant theory, we will concentrate on projective geometry - both synthetic and analytic. We note that this geometry has a surprising range of current applications. Not only do simple problems, such as polyhedral picture recognition, belong to projective geometry [Whiteley 1986], but more subtle problems such as multivariate splines and rigidity of spatial frameworks are also projective [White & Whiteley 1984, Whiteley 1983, 1993].

As the initial language for 'analytic projective geometry' we can take the first-order language of polynomials over the reals (with order). The models are the geometries over real closed fields - with morphisms - the nonsingular linear transformations.

For synthetic projective geometry, we can take a language with geometric 'meets' (intersections), 'joins' (linear extensions), and order. A synthetic projective plane has at least four points, no three collinear, and is closed under forming lines from pairs of distinct points, and constructing points of intersection of points distinct lines. The transformations are *collineations*: maps of points to points, lines to lines, which preserve incidence of a point and a line. For simplicity, we assume that the planes (or higher spaces) satisfy Desargue's Theorem, and Pascal's Theorem, as well as appropriate order axioms to match real closed fields.

3. The Hierarchy of Geometries.

What is 'a geometry'? In a modern restatement of the classical formulation of Felix Klein [1872, 1948], a *geometry* is specified by a category of models and geometric morphisms [Whiteley 1973]. These 'models' also include a first-order language for the semantics of the models. The *geometric properties* are those first-order statements whose truth in a model is *invariant* under the morphisms.

Klein's hierarchy of morphisms runs 'down' from topology, through projective, affine

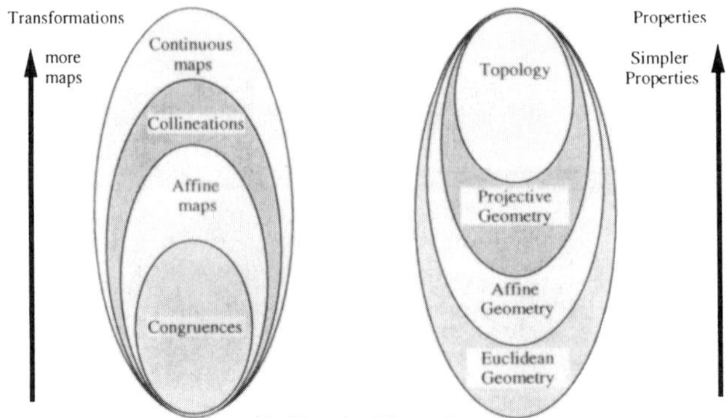

The Hierarchy of Geometries

Figure 1.

and Euclidean geometry (Figure 1). As we move 'up' the hierarchy, we have more geometric morphisms - first congruences, then affine transformations, collineations, and finally all continuous maps. As morphisms increase 'up' the hierarchy, the set of geometric properties decrease and the language for the geometry can be simplified.

This hierarchy has implications for "Learning & Geometry" at three levels: the formal analysis of geometry; human learning; and geometry on computers.

First the formal mathematics. For analytic geometry, more of the underlying algebra is 'geometric' (invariant) as we move 'down' - and the computations become more algebraic. The usual algebra of fields fits Euclidean geometry at the bottom of the hierarchy. The simpler topological properties are more difficult to express in this analytic language.

For synthetic geometry, the constructions become more complex 'down' the hierarchy. The straight-edge constructions of projective geometry are simpler than the ruler, compass and protractor constructions of Euclidean geometry.

If the geometric properties lie higher up the hierarchy (say as projective rather than Euclidean geometry) we should 'simplify' both the language and the geometric methods to be used. Working with the same properties in a lower level of the hierarchy clutters up the presentation with unnecessary details. (A general metatheorem says that invariant theorems have proofs using only these invariants [Whiteley 1977].)

In passing, a further refinement of the hierarchy extracts 'combinatorial' geometric properties: properties which are invariant under 'extensions of the models' [Whiteley 1979]. This semantic (model theoretic) invariance is essentially equivalent to syntactic expression by an open (quantifier-free) formula [Whiteley 1979]. Happily, this syntactic property can be worked into Valiant's theory of the learnable [Valiant 1984].

The implications for human learning are more subtle. It appears that humans learn topology first - then projective, affine and finally Euclidean geometry. (Any two year old child can discover whether a new house is simply connected - or contains a 'race track' around an obstacle.)

A related observation is that 'visual thinking' also concentrates on the simpler properties further up Klein's hierarchy - with Euclidean properties of 'equal distance', 'right angles' etc. being the least 'geometric' [Arnheim 1969]. Venn diagrams, binary trees etc. belong to the very top levels of the hierarchy (see the related analysis of Seligman [1995]). In general, visual thinking eliminates unnecessary details in images – corresponding, in part, to moving up the hierarchy.

Unfortunately, the school curriculum teaches geometry in the opposite order - classically beginning with Euclidean geometry and currently beginning with angles and distances in 'turtle graphics' on computers. In fact, Euclidean geometry has been taught in a fashion that essentially resembles algebra and formal logic. As a result, Euclidean geometry is embedded in the part of the brain associated with sequential, algebraic thinking ('right hand/left brain') not the visual part ('left hand/right brain') associated with topology and shape [Ornstein 1975]. It is no accident that algebra has dominated geometry throughout this century, an imbalance reflected in a general bias towards analytic geometry and the 'bottom' of the hierarchy.

How does this compare with computer geometry? The hierarchy is also reversed in much computational geometry - where objects are represented in the Cartesian coordinates of Euclidean geometry. For example, the Elementary Geometry Theorem Prover of [Chou 1988, Wu 1984] is implemented in Cartesian coordinates. Statements about equal distance are at least as simple as statements about collinearity - and much simpler than statements about connectivity of two objects represented by separate analytic equations.

However in polyhedral picture recognition, the first decisions are topological - and then projective. It is not useful to decide on 'equal distances' etc. when the property sought is projective. Thus effective algorithms begin with combinatorial (topological) properties [Sugihara 1986, Whiteley 1986].

Similarly, an effective designer makes the decisions in order, down the hierarchy from topology to Euclidean geometry [Baracs 1988]. It is inappropriate to decide on Euclidean coordinates for the points before the topological and projective decisions have been made. A computer aided design system *should* respect this order down the hierarchy. It should permit the designer to selectively alter features at all levels of the hierarchy. [On the other hand - current computer graphics and visual display systems have a clear emphasis on approximation to the Euclidean geometry through multiple polygons - a coding that removes all sense of the geometric hierarchy.]

There are good reasons to use the upper levels of the hierarchy for computer geometry. To be effective, this requires other tools than Cartesian coordinates. At a minimum, we need invariant expressions. More strongly, we should maximize the role of synthetic methods - since these are simplified as we move up the hierarchy, while analytic geometry in Cartesian coordinates is not.

At the top of the hierarchy, topology and combinatorics pose an interesting problem. I am not aware of a careful analysis of 'analytic' vs 'synthetic' topology. As topology is taught to undergraduates, it is essentially second-order and analytic: - "for all open neighborhoods ... "!

Compare this with the 'discrete synthetic topology' of graph theory, and its generalization as matroid theory. Many of the formulations for this theory [Nicoletti & White 1986, Brylawski 1986b], have an implicit emphasis on 'synthetic constructions' [Brylawski 1986a]. When matroid theory is done by computers, we have combinatorial (non-numeric) representations and many efficient polynomial algorithms. Thus computational graph theory and computational matroid theory are the basic level of computational synthetic geometry [Bokowski & Sturmfels 1990].

In view of its fundamental position in the hierarchy of geometries, and in human visual thinking, I end this section with a simple challenge:

How can a computer be taught topology?

4. The Equivalence of Synthetic and Analytic Geometry.

The *Fundamental Theorem of Projective Geometry* says that any synthetic plane (or higher space) satisfying Desargues's Theorem and Pascal's Theorem defines a field [Baer 1952]. The statements of synthetic geometry become algebraic statements in the field. The algebraic translations of synthetic properties will automatically be invariant - since the translations of the synthetic transformations include all the geometric transformations of analytic geometry. This implies an injection from synthetic models to analytic models - and from synthetic properties to analytic properties.

Conversely, synthetic constructions can duplicate the algebraic constructions of addition, multiplication, equality etc.. If we restrict ourselves to invariant algebraic statements, with coefficients over the rationals, any first-order analytic statement can be translated (modulo a basis for the projective space) into a first-order synthetic statement [Sturmfels & Whiteley 1991]. In addition, if we work with real projective geometry, the translated algebraic transformations include the geometric transformations, so the translated properties are synthetic geometric properties.

We conclude that, at a formal level:

synthetic projective geometry is equivalent to analytic projective geometry.

In general an equivalence of synthetic and analytic geometries holds for each level of the geometric hierarchy, provided we take care of the precise sets of transformations. (E.g. for projective geometry over the complex numbers, we must include semilinear transformations - with field automorphisms - and not just linear transformations [Whiteley 1991].)

5. The Algorithms for Translation.

We have seen that synthetic and analytic projective geometry are logically equivalent. We have also proposed that synthetic geometry has advantages. What is the connection? These abstract 'translations' hide fundamental issues of complexity. Are there effective algorithms for the translations between analytic and synthetic geometry?

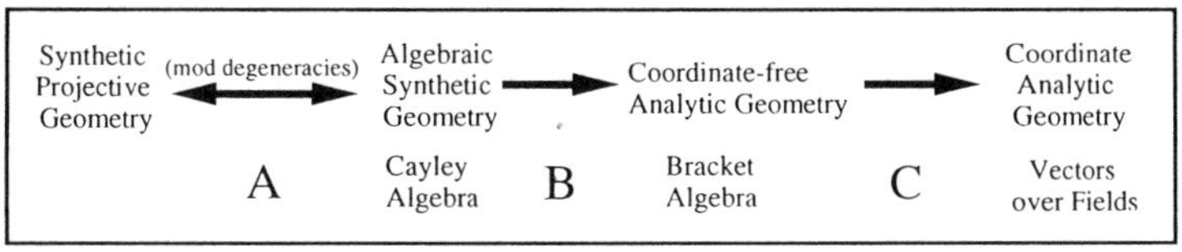

From Synthetic to Analytic Geometry

Figure 2.

¿From synthetic projective geometry, we first move to the simplified language of Rota's Cayley algebra (Grassmann exterior and interior algebra) [Doubilet, Rota, Stein 1976, White 1991, White & Whiteley 1984] (Figure 2A). This is the algebra developed for synthetic geometry - with meets and joins corresponding to geometric intersection and extension - provided that the objects being studied have appropriate scale.

For example, for intersection of two objects, written *meet*: $A \wedge B$, we assume that the points of $A \cup B$ span the entire space (otherwise the answer is zero). For the extension of two objects, written *join*: $A \vee B$, we assume that the flats A and B have no points in common (otherwise the answer is zero). With these assumptions of 'non-degeneracy', Cayley algebra records all of the synthetic statements about finite sets of points, or points, lines, planes etc..

There are simple algorithms for moving from Cayley algebra to polynomial equations in the basic projective invariants - the $n \times n$ determinants for the projective coordinates of points in $(n-1)$-space (Figure 2B). These translations involve some standard operations of 'splits and shuffles' on the sets of points - essentially related to classical identities on the determinants or syzygies of invariant theory [Doubilet, Rota, Stein 1976, White 1991]. The translations are not unique. However, two equivalent translations are connected by the classical syzygies of invariant theory, implemented in the straightening algorithm of invariant theory.

As a convention, we write such determinants as a *bracket* $[a_1 \ldots a_n]$. With the language of rings built on these brackets, and the basic identities of determinants (the syzygies), we have a coordinate-free form for analytic geometry.

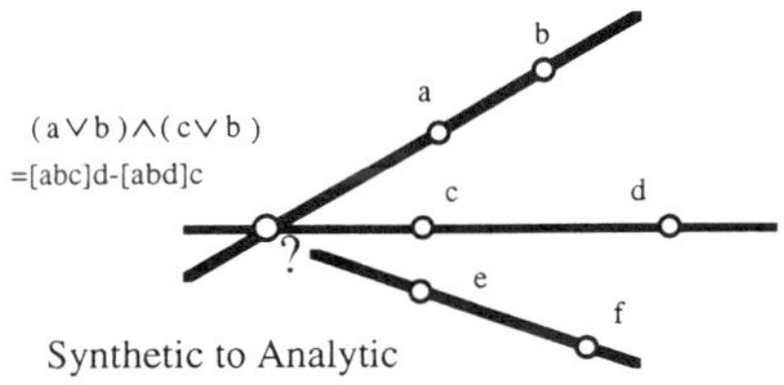

Figure 3.

Example 1. Consider three concurrent lines in the plane (Figure 3):

$$ab \wedge cd \wedge ef = 0$$

The first operation $ab \wedge cd$ yields the point of intersection of the two lines, and the final 0 occurs if and only if this point lies on the third line (or if the first two lines coincided - giving 0 from the degeneracy). This Cayley algebra statement translates to a simple polynomial equation in the brackets:

$$[abc][def] - [abd][cef] = 0.$$

This is clearly zero if the two lines coincide. It is less obvious, but can be checked, that this is zero if and only if the three lines have a projective point in common. ∎

Any polynomial equation in these $n \times n$ determinants can be expanded into the underlying coordinates to reach the standard language of analytic geometry (Figure 2C). Thus we have a direct, constructive translation from synthetic geometry to analytic geometry - for sets of points lines, planes. Of course we increase the number of variables, from 1 per projective point to n for a point in projective $n-1$ space.

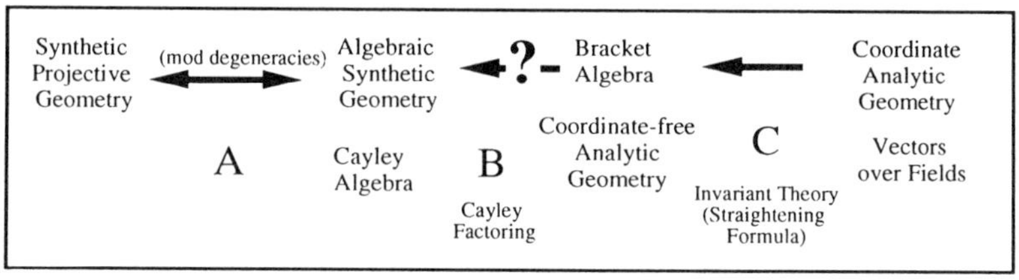

From Analytic to Synthetic Geometry

Figure 4.

It is more difficult to move back from an analytic equation to a synthetic statement (Figure 4). The *First Fundamental Theorem of Invariant Theory* says semantic invariance under transformations can be transformed to syntactic invariance - in the form of writing. Using standard techniques from classical invariant theory, including the straightening algorithm [Doubilet, Rota, Stein 1976, Sturmfels & White 1989], we can translate any invariant formula to an homogeneous formula in the brackets [Whiteley 1991] (Figure 4C). Unfortunately, this straightening algorithm is exponential in complexity.

The essential unsolved algorithmic step is 'factoring' from a polynomial in the brackets back to a synthetic statement: a Cayley algebra statement with just meets and joins (Figure 4B) [White 1991, Sturmfels & Whiteley 1991]. There are simple invariant equations which cannot be factored, as written, into synthetic constructions.

Example 1 continued. The expression: $[abc][def] + [abd][cef] = 0$ does not factor in the Grassmann algebra of the plane. Recall that, in Example 1

$$ab \wedge cd \wedge ef = [abc][def] - [abd][cef].$$

When the sign is changed, this becomes impossible to factor. ∎

There is a more subtle answer. If we multiply this invariant by an appropriate product of brackets, the resulting polynomial does factor as a synthetic construction.

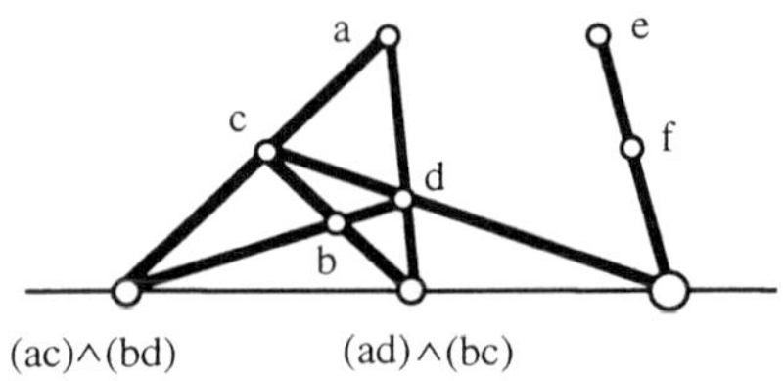

Figure 5.

Example 1 continued. The expression: $[acd][bcd]([abc][def]+[abd][cef])$ does factor, as follows (Figure 5).

$$
\begin{aligned}
\{(ac \wedge bd) \vee (ad \wedge bc)\} \wedge ef \wedge cd \\
&= \{([acb]d - [acd]b) \vee ([adb]c - [adc]b)\} \bigwedge \{[efc]d - [efd]c\} \\
&= \{[acb][adb]dc - [acb][adc]db - [acd][adb]bc\} \bigwedge \{[efc]d - [efd]c\} \\
&= [acb][adc][efd][dbc] - [acd][adb][efc][bcd] \\
&= [acd][bcd]([abc][def] + [abd][cef])
\end{aligned}
$$

This synthetic construction represents the construction of a point of intersection with a cross ratio -1. ∎

Geometrically, this bracket multiplier assumes that certain triples of points span the plane. These non-degeneracy conditions are similar, in spirit, to the 'side conditions' created in Wu's algorithm. However, these bracket monomials are simple to interpret geometrically.

The *Fundamental Theorem of Projective Geometry* implies that every homogeneous bracket polynomial has an appropriate 'premultiplier' so that it factors in the Cayley algebra [Sturmfels & Whiteley 1991]. Unfortunately, this theoretical statement hides an absurdly high power of the monomial - and provides no effective algorithm. Only in the special case of multilinear bracket polynomials is there an effective algorithm [White 1991].

There is a classical unsolved problem which illustrates the complexity of passing from analytic statements to synthetic statements. Over 170 years ago, the Belgian Academy posed the problem of a 3-dimensional analogue of Pascal's Theorem of plane geometry.

Example 2. Six points lie on a conic if and only if

$$\det \begin{bmatrix} a_1^2 & a_2^2 & a_3^2 & a_1a_2 & a_1a_3 & a_2a_3 \\ b_1^2 & b_2^2 & b_3^2 & b_1b_2 & b_1b_3 & b_2b_3 \\ c_1^2 & c_2^2 & c_3^2 & c_1c_2 & c_1c_3 & c_2c_3 \\ d_1^2 & d_2^2 & d_3^2 & d_1d_2 & d_1d_3 & d_2d_3 \\ e_1^2 & e_2^2 & e_3^2 & e_1e_2 & e_1e_3 & e_2e_3 \\ f_1^2 & f_2^2 & f_3^2 & f_1f_2 & f_1f_3 & f_2f_3 \end{bmatrix} = 0$$

This, in turn, translates into a bracket equation, in various forms such as:

$$[abd][bce][fac][def] - [abd][bce][fad][cef] + [abe][bce][fad][cdf] - [abe][bcf][fad][cde] = 0.$$

Notice that this is quadratic in each of the six letters. Finally, this factors as the Cayley algebra equation:

$$\{(ab \wedge de) \vee (bc \wedge ef)\} \wedge (cd \wedge fa) = 0$$

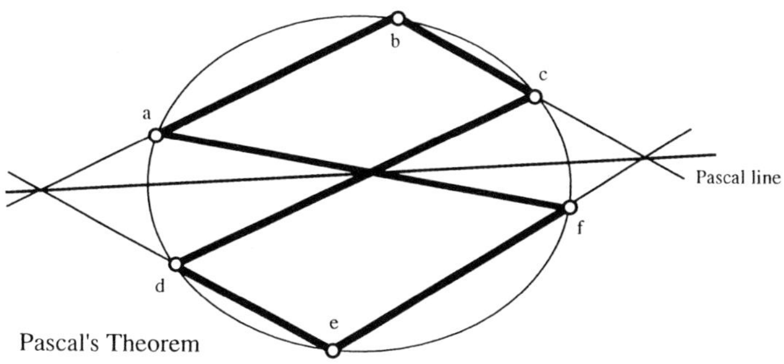

Figure 6.

Translated into synthetic geometry this gives (Figure 6):

Pascal's Theorem. *The three pairs of opposite sides of a hexagon form collinear points (the Pascal line) if and only if the six points lie on a conic section.* ∎

We are now ready for the unsolved problem of a spatial analogue.

Example 3. Consider the simple analytic statement in 3-space:

These ten points lie on a quadric surface.

Equivalently, consider the polynomial equation:

$$det \begin{bmatrix} a_1^2 & a_2^2 & a_3^2 & a_4^2 & a_1a_2 & a_1a_3 & a_1a_4 & a_2a_3 & a_2a_4 & a_3a_4 \\ b_1^2 & b_2^2 & b_3^2 & b_4^2 & b_1b_2 & b_1b_3 & b_1b_4 & b_2b_3 & b_2b_4 & b_3b_4 \\ c_1^2 & c_2^2 & c_3^2 & c_4^2 & c_1c_2 & c_1c_3 & c_1c_4 & c_2c_3 & c_2c_4 & c_3c_4 \\ d_1^2 & d_2^2 & d_3^2 & d_4^2 & d_1d_2 & d_1d_3 & d_1d_4 & d_2d_3 & d_2d_4 & d_3d_4 \\ e_1^2 & e_2^2 & e_3^2 & e_4^2 & e_1e_2 & e_1e_3 & e_1e_4 & e_2e_3 & e_2e_4 & e_3e_4 \\ f_1^2 & f_2^2 & f_3^2 & f_4^2 & f_1f_2 & f_1f_3 & f_1f_4 & f_2f_3 & f_2f_4 & f_3f_4 \\ g_1^2 & g_2^2 & g_3^2 & g_4^2 & g_1g_2 & g_1g_3 & g_1g_4 & g_2g_3 & g_2g_4 & g_3g_4 \\ h_1^2 & h_2^2 & h_3^2 & h_4^2 & h_1h_2 & h_1h_3 & h_1h_4 & h_2h_3 & h_2h_4 & h3h4 \\ i_1^2 & i_2^2 & i_3^2 & i_4^2 & i_1i_2 & i_1i_3 & i_1i_4 & i_2i_3 & i_2i_4 & i_3i_4 \\ j_1^2 & j_2^2 & j_3^2 & j_4^2 & j_1j_2 & j_1j_3 & j_1j_4 & j_2j_3 & j_2j_4 & j_3j_4 \end{bmatrix} = 0$$

This invariant polynomial equation can be rewritten as a bracket polynomial, which has been extensively studied [Turnbull & Young 1926]. Recently, White [1991] straightened the polynomial to a sum of 138 monomials of degree 5 in the brackets, with coefficients ± 1. However, in this form, it does not factor into the Cayley algebra. A translation of the unsolved problem of the Belgian Academy is (see also [Chasles 1875]):

Find a small monomial so that this expression factors in the Cayley Algebra.

The general theorem of [Sturmfels & Whiteley 1991] gives an upper bound for the degree of a factoring monomial of approximately $100,000$!

On the other hand, sufficient conditions are known [Sennet 1876]:

A skew decagon is inscribable in a surface of second order if its opposite sides intersect (in pairs) in points on a single plane.

Each of these coplanarities is a quadratic equation which factors without a multiplier. Unfortunately, we have at least three distinct equations, so the conditions are far from necessary. ∎

The difficulties in 'translation' back to synthetic geometry suggest trying to keep the calculation as 'synthetic' as possible - to minimize the use of exponential (or unknown) translation algorithms. If the algorithms used for geometric theorem proving are going to generate extra expressions (side conditions), these should be connected to the geometry. If the initial property is written in Cartesian coordinates, this connection is a non-trivial task.

Recent work in computer vision has pointed out other advantages of invariant representation of the basic objects [Forsyth et al 1990]. In the long term, we propose truly synthetic computations for synthetic properties [Havel et al 1988].

6. The Role of Identities in Analytic Geometric Proofs.

We have promoted the use of synthetic methods and invariant expressions for computer geometry. However analytic geometry will continue to play an important role in computer geometry. We have a specific proposal about how analytic proofs should be recorded, for both Cartesian and invariant presentations.

In a formal first-order language for analytic geometry, every open formula can be written in conjunctive normal form,

$$(f_1 \neq 0 \vee \ldots \vee f_m \neq 0 \bigvee g_1 = 0 \vee \ldots \vee g_n = 0) \& \ldots \& (f_p \neq 0 \vee \ldots \vee g_s = 0).$$

Equivalently, every such formula can be written:

$$(f_1 = 0 \& \ldots \& f_k = 0 \Rightarrow g_1 = 0 \vee \ldots \vee g_n = 0) \ \& \ldots \& \ (\ldots \& f_m = 0 \Rightarrow \vee g_r = 0 \ldots).$$

If we are working with an appropriate theory for real closed fields, the truth of each conjunct is – via the 'Realstellensatz'– equivalent to a single polynomial equation, which contains all the same information (and, apparently more see [Whiteley 1991]). For example:

A formula of the form:

$$f_1 = 0 \& f_2 = 0 \& \ldots \& f_k = 0 \Rightarrow g_1 = 0 \vee \ldots \vee g_m = 0,$$

is true over a real closed field if and only if there are polynomials a_i, b_k, with integer coefficients, and positive integers n and m such that:

$$\sum_i a_i f_i = m \prod_j (g_j)^{2n} + \sum_k (b_k)^2.$$

The slogan here is:

A logical implication yields an algebraic implication.

It is well understood, in most forms of analytic geometry, that we can prove such a 'Theorem' by deriving identities. If the theorem holds over algebraically closed fields, we have the simpler pattern of Hilbert's Nullstellensatz:

A formula of the form:

$$f_1 = 0 \& f_2 = 0 \& \ldots \& f_k = 0 \Rightarrow g_1 = 0 \vee \ldots \vee g_m = 0$$

where f_i and g_j are polynomials with integer coefficients, is true over an algebraically closed field if and only if there are polynomials a_i, with coefficients in the integers, an integer $k \neq 0$ in the field, and integers n_j such that:

$$\sum_i a_i f_i = k \prod_j (g_j^{n_j})$$

as polynomials.

Of course if the properties in the theorem are invariant, all of the added polynomials can also be made invariant. For example, a theorem in the language of brackets has an identity, and a proof, within this coordinate-free language.

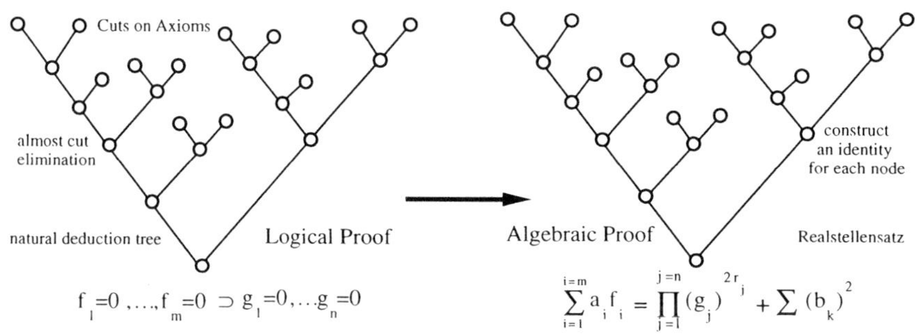

Figure 7.

I want to promote a converse. A proof theoretic argument [Whiteley 1991] shows that:

Any natural first-order proof of such a disjunction can be read by a constructive algorithm to give the corresponding identity.

Paraphrased (Figure 7):

the Nullstellensatz and Realstellensatz identities grow on natural proof trees.

If we are going to work with analytic geometry on computers, then we should arrange the algorithms so that they record the corresponding identity for each conjunct of its statements [see the next section]. In particular, conjunctive normal forms of theorems correspond to conjunctions of identities. Conveniently, this is just right for Valiant's theory of the learnable.

Example 4. Consider the problem of coordinatizing a matroid of rank n over the complex numbers. Such a configuration has certain n-tuples independent, and other n-tuples dependent. Thus a set of coordinates must satisfy a first order formula:

$$(\exists \mathbf{x}_1 \dots \mathbf{x}_k)([\mathbf{x}_1 \dots \mathbf{x}_n] = 0 \& \dots \& [\mathbf{x}_s \dots \mathbf{x}_t] \neq 0 \& \dots).$$

A proof of non-coordinatizability over the complex numbers (or over any extension of the rationals) is therefore a theorem of the form:

$$\neg(\exists \mathbf{x}_1 \dots \mathbf{x}_k)\neg([\mathbf{x}_1 \dots \mathbf{x}_n] = 0 \& \dots \& [\mathbf{x}_s \dots \mathbf{x}_t] \neq 0 \& \dots)$$

or $$(\forall \mathbf{x}_1 \dots \mathbf{x}_k)([\mathbf{x}_1 \dots \mathbf{x}_n] \neq 0 \vee \dots [\mathbf{x}_j \dots \mathbf{x}_m] \neq 0 \vee [\mathbf{x}_s \dots \mathbf{x}_t] = 0 \vee \dots).$$

This universal theorem is equivalent to a single Nullstellensatz identity:

$$a[\mathbf{x}_1 \dots \mathbf{x}_n] + \dots + c[\mathbf{x}_j \dots \mathbf{x}_m] = [\mathbf{x}_s \dots \mathbf{x}_t]^p \dots [\mathbf{x}_d \dots \mathbf{x}_r]^q.$$

In their studies of coordinatizing matroids, Bokowski and Sturmfels [1990] have called this Nullstellensatz identity the *final polynomial* of a proof of non-coordinatizability.

7. 'Better' Algebraic Proofs for Geometry.

In a context where alternate algebraic algorithms and implimentations can generate 'proofs' for the same theorem of elementry geometry, it is important to compare the 'quality' of the proofs, in addition to the speed at which the proofs are derived. What makes one proof in geometry 'better' than another proof? I conclude this essay with some specific criteria for comparing alternate proofs of geometric theorems produced by computerised algebraic algorithms.

Of course, the primary criterion is:

(i) *Will the algorithm generate a proof, within a reasonable time?*

In particular, if the hypotheses are incomplete, a number of these algorithms generate additional 'side conditions' (typically assumptions of non-degeneracy) which are sufficient for a proof. (This is an improvement over returning no proof.)

(ia) *Will the algorithm complete the non-degeneracy conditions and generate a proof?*

For important classes of 'elementary geometry' theorems, a number of competing algorithms pass this test. Many seem to have comparable times, in practice. (For a practicing geometer the difference between 1 second and 10 seconds is not important, although 10 minutes is better than 100 minutes.)

There are a number of other reasons to prefer one proof to another, given that both are correct. Some obvious issues are the 'quality' of the theorem and the 'quality' of the proof.

The additional 'side conditions' are an artifact of the specific proof technique implimented in the algorithm, or of the translation between the geometric statement and the algebraic formulae used in the input. They may not be necessary conditions for the truth of the conclusion.

(ii) *Are all (added) hypothesis necessary?*

This is particularly evident in algorithms (or implementations) which begin with a 'construction' of the configuration. In most cases, there are several alternate sequences for the construction, and each construction sequence leads to its own set of non-degeneracy conditions. Very often these include inessential conditions. Even if the particular set is minimal (no hypothesis can be removed), there may be alternate minimal completions of the hypotheses. If a method systematically generates inessential non-degeneracy conditions, this should be explicitly noted.

Sometimes the added hypotheses (polynomial equations) do not have 'geometric interpretations'. In terms of the classical foundations of analytic geometry, these equations are *not invariant* for the transformations of the corresponding geometry [Whiteley 1991]. Such an event is clearly an artifact of the proof technique.

(iii) *Do all added hypotheses lie at the level of 'geometry' in the original hypothesese and conclusion?*

(iiia) *Can all added (algebraic) hypotheses be translated into synthetic geometric conditions?*

The previous section on the identities for algebraic proofs suggests several additional criteria.

(iv) *Does the proof apply to fields of arbitrary characteristic?*

(iva) *Does the proof apply only to the reals, or to all fields of characteristic zero?*

(v) *Does the proof yield a simple (low degree) Nullstellensatz identity for the theorem?*

Given a Nulltstellensatz identity, however derived, it is a simple matter to verify that the proof is correct, and that it proves the stated theorem. This is a specific form of a more general criterion:

(vi) *Is it easy to verify the correctness of the proof from the output?*

I make no claim that these criteria are complete or final. Other justifications have been offered to support alternate mechanical derivations of existing theorems, including:

(vii) their similarity to 'human generated' patterns - and hence their pedagogical value;

(viii) the techniques extend to higher dimensions;
(ix) the techniques suggest new, original results.

Essentially, I am convinced that alternate methods should be compared. I believe that certain logical, geometric and algebraic fundamentals play an important role in this comparison. My goal is to stimulate a broad agreement on basic criteria, both to evaluate existing methods, and to develope additional methods which are 'better' by these criteria.

Finally, I note that essential unsolved problems remain, such as translating from the anlaytic language of brackets to synthetic constructions. The companion paper [Whiteley 1995] presents some other unsolved geometric problems related to representing and manipulating geometric properties on computers.

Bibliography

[Arnheim 1969]

R. Arnheim, **Visual Thinking**, University of California Press, Berkeley, Los Angeles, London.

[Baracs 1988]

J. Baracs, Spatial perception and creativity; in **Shaping Space: A Polyhedral Approach**, (M. Senechal and G. Fleck eds.), Birkhauser, Boston, 118-132.

[Baer 1952]

R. Baer, **Linear Algebra and Projective Geometry**, Academic Press, New York.

[Bokowski & Sturmfels 1990]

J. Bokowski and B. Sturmfels, **Computational Synthetic Geometry**, Lecture Notes in Mathematics 1355, Springer-Verlag, Berlin, Heidelberg, New York.

[Brylawski 1986a]

T. Brylawski, Constructions; in **Theory of Matroids**, (N. White ed.), Encyclopedia of Mathematics, Cambridge University Press, New York, 127-223.

[Brylawski 1986b]

T. Brylawski, Appendix of matroid cryptomorphisms; in **Theory of Matroids**, (N. White ed.), Encyclopedia of Mathematics, Cambridge University Press, New York, 298-312.

[Chasles 1875]

M. Chasles, **Aperçu Historique des Méthodes en Geometrie**, Paris 1875.

[Chou 1988]

S-C. Chou, **Mechanical Geometry Theorem Proving**, Mathematics and its applications Vol 41, D. Reidel Pub. Co.; Norwell, MA, U.S.A.

[Doubilet, Rota, Stein 1976]

P. Doubilet, G-C Rota and J. Stein, On the foundations of combinatorial theory IX: Combinatorial Methods in Invariant Theory, *Studies in Applied Math.* **56**, 185-216.

[Forsyth et al 1990]

D. Forsyth, J. Mundy and A. Zisserman, Applications of invariant theory in computer vision, preprint, Robotics Research Group, Department of Engineering Science, Oxford University, England.

[Havel et al 1989]

T. Havel, B. Sturmfels and N. White, Proposal for a geometric algebra software package, *SIGSAM Bulletin* **23**, p. 13-15

[Klein 1872]

F. Klein, **Vergleichende Betrachtungen über neure geometrische** Forschungen, Erlangen, (*Math. Annalen* 43, p63-).

[Klein 1948]

F. Klein, **Elementary Mathematics from an Advanced Standpoint: Geometry**, Dover, New York.

[Nicoletti & White 1986]

G. Nicoletti and N. White, Axiom systems; in **Theory of Matroids**, (N. White ed.), Encyclopedia of Mathematics, Cambridge University Press, New York, 29-44.

[Ornstein 1975]

R. Ornstein, **The Psychology of Conciousness**, Penguin, New York.

[Sennet 1876]

P. Sennet, Note sur une class particulaire de décagones gauches inscriptable à l'ellipsöide, *Comptes Rendus (Paris)* **82**, 162-165.

[Seligman 1995]

J. Seligman, An algebraic appreciation of diagrams, manuscript, Philosophy Dept., Indiana University. Available by anonymous ftp from phil.indiana.edu as /pub/Seligman/aad.ps.z.

[Sugihara 1986]

K. Sugihara, **Machine Interpretation of Line Drawings**, MIT Press Cambridge MA.

[Sturmfels & White 1989]

B. Sturmfels and N. White, Gröbner bases and invariant theory, *Advances in Mathematics* 76, 245-259.

[Sturmfels & Whiteley 1991]

B. Sturmfels and W. Whiteley, On the synthetic factorization of projectively invariant polynomials, *J. Symbolic Computation* 11, Special Issue on Invariant Theoretic Methods in Symbolic Computational Geometry, 439-453.

[Turnbull & Young 1926]

H.W. Turnbull and A. Young, The linear invariants of ten quaternary quadrics, *Trans. Cambridge Puil. Soc.* **23**, 265-301.

[Valiant 1984]

L.G. Valiant, A theory of the learnable, *Comm. of the ACM* **Vol. 27**, 1134-1142.

[White 1991]

N. White, Cayley factorization of multilinear bracket polynomials, *J. Symbolic Computation* 11, Special Issue on Invariant Theoretic Methods in Symbolic Computational Geometry, 421-438.

[White & Whiteley 1984]

N. White and W. Whiteley, The algebraic geometry of stresses in frameworks, *SIAM Journal of Algebraic and Discrete Methods* 4, 481-511.

[Whiteley 1973]

W. Whiteley, Logic and invariant theory I: invariant theory of projective spaces, *Trans. Amer. Math. Soc.* 177, 121-139.

[Whiteley 1977]

______, Logic and invariant theory III. Axiom systems and syzygies, *J. London Math. Soc. (2)* **115**, 1-15.

[Whiteley 1979]

______, Logic and invariant theory IV, invariants and syzygies in combinatorial geometry, *J. Combinatorial Theory* B 26, 251-267.

[Whiteley 1986],

______, Two algorithms for polyhedral pictures, *Proceedings of the Second ACM Conference on Computational Geometry*, 142-149.

[Whiteley 1991]

W. Whiteley, Logic and invariant computation for analytic geometry, *J. Symbolic Computation* 11, Special Issue on Invariant Theoretic Methods in Symbolic Computational Geometry, 549-578.

[Whiteley 1993]

______, Matroids and rigidity, in **Matroid Applications**, Neil White (ed.), Cambridge University Press, to appear.

[Whiteley 1995]

______, Representing Geometric Configurations, this volume.

[Wu 1984]

Wu Wen-tsün, Basic principles of mechanical theorem-proving in elementary geometries, *J. of Sys. Science. & Math. Sciences.* 4, 207-235.

Representing Geometric Configurations

Walter Whiteley§
Department of Mathematics and Statistics
York University
4700 Keele St, North York, Ontario, M3J-1P3, Canada
Whiteley@mathstat.YorkU.ca

Abstract.

How should we represent a geometric configuration? Building on artificial intelligence investigations of 'knowledge representation' and research on constrained diagrams in CAD, criteria are presented for an 'adequate representation' of a discrete geometric set. Dependence on both the level of geometry (Euclidean, similarity, and projective) and on the sources of the geometric data are considered. The criteria are illustrated with representations for a finite set of 'geometric points and lines' in the plane. Appropriate constructions and representations are translated from classical and from modern geometric practice. The sample representations use coordinates, distances, angles, incidences, orientations and projections.

1. Introduction

We will examine the problem of recording a discrete geometric configuration drawing on three points of view:

(A) the search for 'geometric constructions' and 'unique specification' of geometric configurations in many areas of geometric practice;

(B) Klein's characterization of geometries by invariance under a group of transformations;

(C) the search for 'adequate knowledge representations' for problem solving in artificial intelligence.

What are the connections among these approaches?

---

§ This work was supported by grants from FCAR (Québec) and NSERC (Canada).

(A) In everyday geometric work, we search for and use 'adequate representations' of geometric objects. We list a few examples which we will draw on throughout the paper.

(a) In analytic geometry, Cartesian coordinates represent the points, lines, planes etc..

(b) In synthetic geometry, a configuration is described by a construction (joining points with a line, drawing a perpendicular line, bisecting a line segment or an angle, etc.). In a 'dynamic geometry program' (such as Geometer's Sketchpad or Cabri) choices made during this construction (such as selecting a point on a line) can then be animated to generate a range of final objects for the range of choices.

(c) An engineer with a CAD program represents an object by a sketched 'drawing' and a series of dimensions (distances, angles etc.) which specify a unique object [Sugihara 1985, Owen 1991, 1993].

(d) A draftsman represents a spatial object as two (or more) related orthogonal plane projections.

(e) A designer may represent a pattern by selecting the combinatorial and topological pattern (vertices, edges, connectivity), then making further choices of incidences, parallels, distances and angles until there are no more choices and the object is 'determined' [Baracs 1988, Whiteley 1994].

(f) A chemist measures a molecule by a set of inter-atomic distances from nuclear magnetic resonance (NMR) data [Crippen & Havel 1988].

(g) A civil engineer represents a structure as a framework of joints and bars - to determine if the position is unique - or flexible [Asimow & Roth 1978, Graver et al 1993, Connelly 1990].

(h) A surveyor measures distances and directions (angles) in order to prepare a map [Snay & Spriggs 1982].

(i) A 'computational geometer' chooses a representation to organize data bases and to speed up computations. If the representation is affine or projective, then additional techniques, such a polarity, projection, or section, will be available.

To use geometric representations well is to follow paths marked by the geometric masters. The sources for our examples in this paper and the companion paper [Whiteley 1994] include: Euclid, Cauchy, Maxwell, Cremona, Hilbert, Steinitz, Dehn, Alexandrov .... The theorems, examples, and unsolved problems originate in scattered fields of ancient and modern research, such as ruler and compass constructions, projective geometry, uniqueness of convex polyhedra, rigidity of frameworks, and analysis of polyhedral pictures. Although the original work was not expressed as 'a representation', this 'representations perspective' unites the fundamental geometric patterns - and raises interesting new geometric problems.

In modern geometry, representations are often vectors of numerical values for selected features, and an algebra for the relevant constructions in terms of these numbers. There are other possibilities - such as the combinatorial patterns and matroid methods [White 1986], or straight-edge constructions used in synthetic geometry. We will concentrate on the numerical and combinatorial representations of finite configurations in the plane. The features we examine will include: Cartesian coordinates; distances between points; angles between lines; angles between planes; incidences of points, lines and planes; projections of points; and orientations of simplices.

(B) A first step in selecting a representation is to decide the 'level of geometry'. As Felix Klein has taught us, the 'level of geometry' is recorded by a group of transformations (see [Klein 1872]). We must ask: What transformations of the objects match a single representative? Conversely, if we select a group of transformations, what representatives will identify unique equivalence classes of objects?

Two representations are 'equivalent' if they fit the same group of transformations. We will present several representations for plane Euclidean geometry (the congruence maps), for similarity maps (congruences and dilations), and for projective geometry.

If a problem lies at a particular level of geometry, such as projective geometry, then its representation, and solution, should lie at this level as well. Of course 'simple features' (incidences from projective geometry) can also be used at 'complex levels', such as Euclidean geometry. However, complex features, such as distances, are misleading and irrelevant at simple levels such as projective geometry [Forsyth et al 1990].

Some constructions and representations mix features from several levels, working down from the simple to the more complex. In particular, it is appropriate for a designer to work 'down the hierarchy', selecting features at each level until the representation (and therefore the design) is complete [Baracs 1988].

(C) For a fixed group of transformations, there are many representations. How do we select one? How and why do we change from one to another?

Some of these issues have been studied in artificial intelligence (AI) [Bench-Capon 1990]. However the problems of 'adequate knowledge representation' are far broader than those of artificial intelligence. Our examples are based on current geometrical experiences. After a few introductory examples (Sections 2,3), we propose a modified list of criteria for 'adequate geometric representations' (Section 4). We adapt criteria offered in artificial intelligence, to extend the applications to other areas of geometric practice, and to draw out the particular characteristics of geometry.

In Sections 5-7, these criteria and their connection to geometric problems are illustrated with three representations of plane configurations:

(a) representation, up to congruence, by distances between pairs of points, (Section 5) (drawn from structural rigidity and molecular chemistry);

(b) representation, up to similarity, by lines and angles between lines (Section 6) (drawn from surveying and engineering drawings);

(c) representation, up to a projective map, by incidences of points and lines (Section 7) (drawn from synthetic geometry).

For 3-space, the general problems of representation of an arbitrary set of points is too hard. In an companion paper [Whiteley 1994], we summerize old and new results for a special class of geometric objects: the vertices, edges and faces of 'spherical polyhedra'.

Our analysis carries two messages. For users of representations, we highlight the importance of good geometry. A large collection of geometric studies implicitly seek a 'good representation' and this search is a unifying theme for many deep geometric results. To use geometric representations well, we must know geometry and not just use a geometric language.

For geometers, the search for geometric representations raises significant geometric problems. Within fields as disparate as structural engineering, CAD and artificial intelligence, 'representation' offers a direction for future geometric research. The perspective of 'representation' illuminates common themes and suggests important analogies to solve current geometric problems.

2. A Basic Example: Cartesian Coordinates for Points in the Plane.

Consider a finite set of labelled geometric points in the plane (Figure 1A). This is a typical, if very simple, plane configuration. (A single picture of the points can serve as an initial representation.)

The *Cartesian representation* of a plane geometric set is:

(i) an indexing set V of *vertices* for the points;

(ii) a *location map* $\mathbf{p} : V \to I\!R^2$ which gives plane coordinates $\mathbf{p}(v_i) = \mathbf{p}_i = (x_i, y_i)$ for the vertices (Figure 1B). We assume that two indices may have the same location.

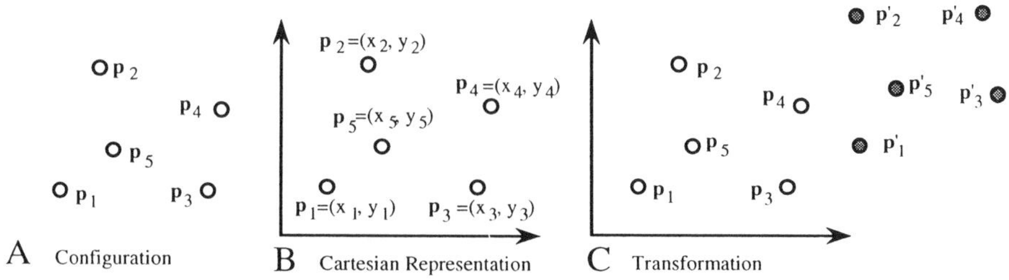

Figure 1.

This representative is extracted from the picture by selecting two perpendicular lines as axes, a scale for each axis, and measuring the distances from these lines. This Cartesian representation is used in most analytic geometry and in computer programs for geometry.

When are two sets of plane points 'geometrically equivalent' (Figure 1C)? Following [Klein 1872, 1948], a geometry is characterized by a model and a group of transformations. Two configurations are *geometrically equivalent* if some transformation takes one to the other. For example, Euclidean geometry is characterized by the congruences (translations, rotations and reflections). Projective geometry is characterized by the collineations, which preserve lines and incidences. Topology is characterized by continuous transformations of the entire plane.

For a finite configuration, there are other transformations which are restricted to these points and to selected properties. Thus, we may preserve only collinearity among these points, and ignore additional intersections of lines through these points. This creates the geometry of a matroid [White 1986]. Or we could preserve these lines plus the orientations of triangles - creating the geometry of an oriented matroid.

As basic examples, we will concentrate on three types of geometry: Euclidean geometry, similarity (which adds dilations or similarity maps) and projective geometry.

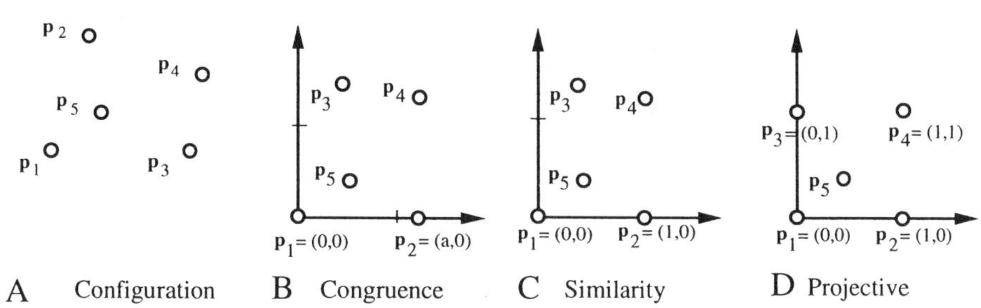

Figure 2.

For *congruence maps* (translations, rotations and reflections) the Cartesian representative is not unique. Each picture has many different representatives, corresponding to the selection of the axes. The following transformation gives a unique location relative to the axes (Figure 2B):

(a) *place vertex 1 at the origin;*

(b) *place the next distinct vertex on the positive x-axis, preserving the distance from vertex 1;*

(c) *place the next vertex not collinear with these two on the positive side of the x-axis; preserving the distances to the previous two vertices.*

This selects a unique Cartesian representative for the geometric set, under the congruence maps.

Remark 1. A geometric transformation has two distinct interpretations - as an 'alias' and as an 'alibi'. The *alias* interpretation says we have different names for a fixed configuration: we can choose the origin and the coordinate axes where we wish - giving different coordinates to the same picture (change the names not the picture). The *alibi* interpretation says that we move one configuration to another (change the picture to give new locations to the points). Our figures, and our vocabulary, are based on the alibi interpretation.

For similarity maps (congruences and dilations), step (b) changes to:

(b') *place the next distinct vertex at* $(1, 0)$.

This selects a unique location for each geometric set (Figure 2C), under the similarity maps.

Projective maps also require:

(d''') *place the next vertex not collinear with any pair of the previous vertices at the point* $(1, 1)$*; otherwise, find the first vertex collinear with two other designated points and place it at the midpoint of their line segment.*

This selects a unique location for each geometric set (Figure 2D), under the projective maps.

Analytic geometers and physicists employ such a 'selection of origin and axes' to simplify their calculations. This selection is also embedded in computer systems such as the Elementary Geometry Theorem Prover (see [Wu 1984, Chou 1988]).

Within a 'coordinate system', we have formulae for calculating other 'geometric properties': distances between points; angles between edges; collinearity of three vertices; concurrence of three lines; etc.. For a large space of transformations, such as the projective transformations, many formulae, such as distances between points, are not 'invariant'. These 'variant' formulae do not record 'geometric properties' of the underlying geometric points. Extracting invariant formulae (or equations) is the 'First Fundamental Problem' of the invariant theory for an analytic geometry. This leads to a 'coordinate free' representation of the geometric configurations (see [Whiteley 1991]).

There are other traditional coordinates for points in the plane. For example, 'homogeneous coordinates' are used for projective geometry. For Euclidean geometry, we also have polar coordinates. We will not pursue these general alternatives here.

3. Constructions and Representations

Synthetic Euclidean geometry 'represents' a configuration by a *straight-edge construction*. The idea is simple. Consider a set of points and lines generated by the following steps:

(i) *Choose a point not on any previous line;*
(ii) *Choose a line not though any previous point;*
(iii) *Choose a line though a given point;*
(iv) *Choose a point on a given line;*
(v) *Draw the line though two points;*
(vi) *Draw the point of intersection of two lines.*

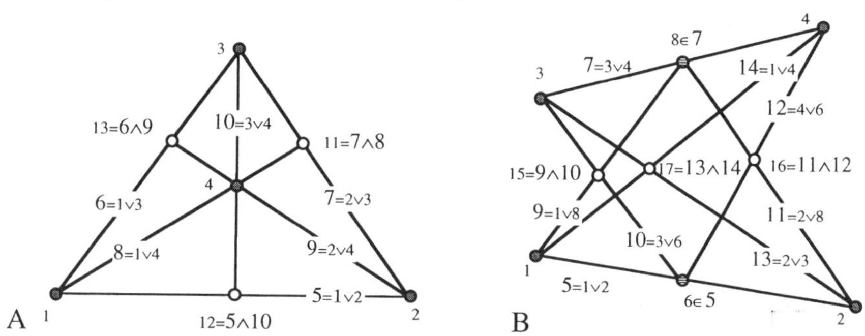

Figure 3.

Example 1. Figure 3A gives a numbered sequence of points and lines. The points are small print and lines are large print. The notation $1 \vee 2$ means the line through points 1 and 2, $1 \wedge 2$ means the point of intersection of lines 1 and 2, and $5 \in 6$ means point 5 is on line 6. This construction is a projective representation: after choosing four points, no three on a line (four uses of step 1 shown by shaded points), all other points are uniquely defined. From the 'special location' of Section 2, for projective maps, linear algebra finds the Cartesian coordinates for the remaining points and lines.

Figure 3B shows a second construction. This is not a complete projective representation, since, after placing the first four points, the construction chooses two more points on existing lines (lined circles). To complete this 'partial representation' we must specify these 'choices' by additional data. ∎

Each straight-edge construction yields at least a partial representation of the configuration. The construction may leave a set of extra 'choices', such as slopes of lines, or ratios for three points on a line, and the Cartesian coordinates of all the points and lines are rational functions of these choices. (In fact, if the coordinates of the original points, and of these choices, are rational numbers, all of the points and lines have rational coordinates.)

Starting from a 'partial representation' arising in a construction, additional incidence conditions might reduce the choices, and produce a full representation.

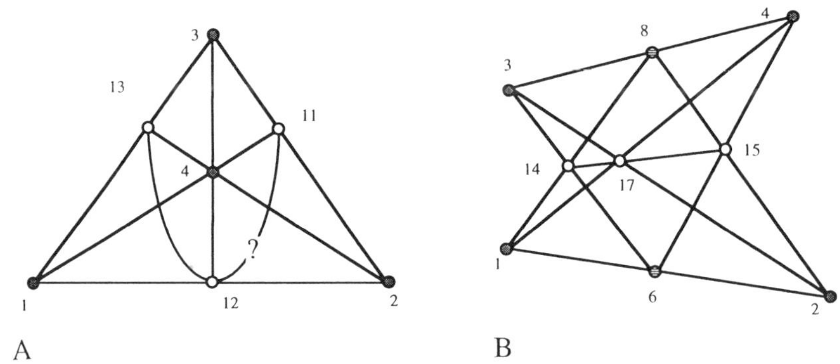

Figure 4.

Example 2. Consider the construction of Figure 4A. The additional condition: that the line $11 \vee 12$ contains point 13 (Figure 4B), creates a geometric problem. The pattern cannot be found in the plane [Hilbert & Cohn-Vossen 1956, Bokowski & Sturmfels 1990]. (In algebraic terms, this 'Fano plane' requires a field of characteristic 2.) We have given information beyond a full representation, and it is inconsistent.

Consider the construction of Figure 3B. The additional requirement that 14, 15 and 16 are collinear (Figure 4B), might reduce the choices. In fact this collinearity is

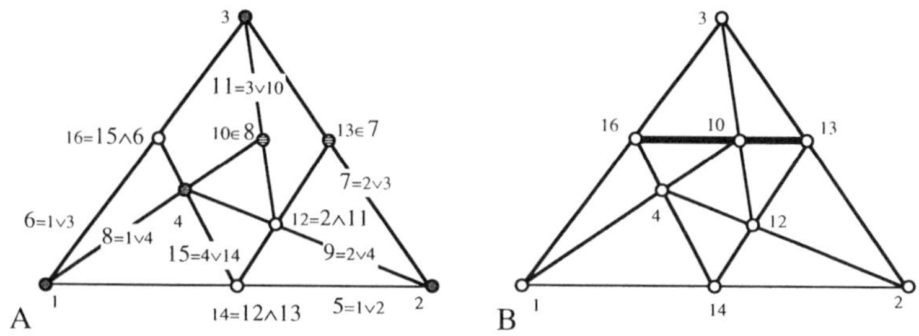

Figure 5.

guaranteed by Pappus' Theorem of projective geometry, so we have changed nothing! This 'extra data' was redundant - even for the partial representation.

Figure 5A indicates another construction of nine points - also with two 'choices' (one for point 10 on its line, and one for point 13 on its line). The condition that 10, 13, 16 are collinear (Figure 5B) is realizable - and removes one choice. This is a partial representation, but we have no obvious 'construction' for the correct choice of point 13. ∎

Certain additional incidences will complete a partial representation to a complete representation. We return to the study of these incidence representations in Section 7.

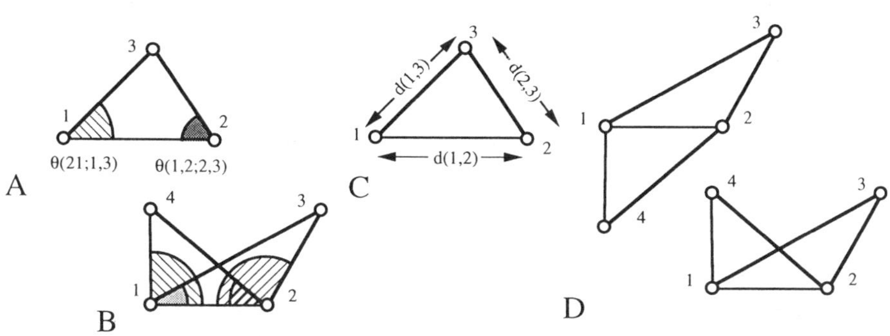

Figure 6.

We can also 'construct' a Euclidean configuration with ruler, compass and protractor.

Example 3. The triangle can be constructed, up to similarity, by two angles (Figure 6A). We can fix four points, up to similarity, by four angles (with signs) (Figure 6B).

A triangle in the plane can also be 'constructed', up to congruence, from three distances (Figure 6C). If we use five distances for four points, there are, up to congruence, two distinct patterns (Figure 6D). The distances now give a local, but not a global representation. ∎

Not all representations give 'constructions' of the Cartesian location of the points.

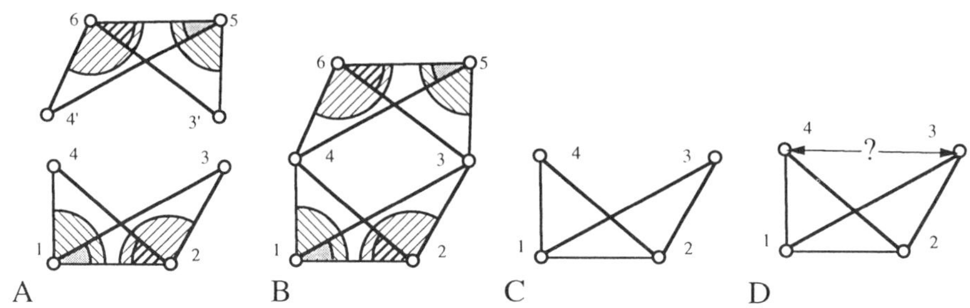

Figure 7.

Example 4. If we match up two of the 'four angle' configurations (Figure 7A,B), there is a unique configuration of six points. This is a representation of the points (and lines). However this data does not give an immediate formula for the Cartesian coordinates. Algebraically, we must solve a system of non-linear equations.

If we give the lengths of the edges of a piece of this configuration (Figure 7C), there is a compass and ruler construction of the configuration, up to a reflection. The reflection is removed by an extra distance $d(3,4)$ (Figure 7D): but there are only two discrete choices for this distance. Most 'chosen' values of $d(3,4)$ will have no location. If there is a realization for the 6 distances, it will be unique. ∎

Section 5 examines distances as representations up to congruence and Section 6 presents angles as representations up to similarity. So far, our basic message is 'good representations' are scarce and real world representations present serious difficulties.

4. Good Representations for Geometric Configurations.

What makes a 'good representation' of a geometric object? The previous examples illustrates some, but not all, of the issues. We have already seen that the transformations of the geometry influence the choice of features.

The representation is also determined by the tools used to measure or present the original object. If a plane configuration is presented as a surveyor's set of distances and angles, we will not start with the Cartesian representative or a plane picture. A computer program to design polyhedral shapes should not ask for input in Cartesian coordinates. Instead, it should accept design decisions in a natural order, beginning with combinatorial decisions (how many faces, what topology) and then geometric decisions about incidences, parallels, and finally lengths etc. [Baracs 1988]. We must be able to recognize an 'adequate representation' in these other geometric vocabularies.

For computer problem solving in artificial intelligence, Bench-Capon [1990] presents criteria for an 'adequate representation of knowledge'. I modify these criteria to reflect both our restriction to geometries and our concern with applications beyond artificial intelligence.

(i) *Each representative should fit some object, and every object should have some representation.* It should not contradict the axioms of the geometry. (This corresponds to Bench-Capon's 'metaphysical adequacy'.)

(ii) *A representative should fit a 'unique object'.* If two objects have the same representation - they should be 'equivalent' for the geometry. Otherwise the representation should be extended to distinguish the objects. (This corresponds to Bench-Capon's 'epistemic adequacy'. It also reflects his 'lack of ambiguity'.)

(iii) *A class of geometrically equivalent objects should have a single representative.* Otherwise we need a clear set of transformations which carries one representative into any other representative. (This is related to Bench-Capon's criterion of 'relevance', since features which are irrelevant for the geometry will violate this standard.)

(iv) *The representation should be minimal (non-redundant).* If we delete one element, some pair of non-equivalent objects should drop to the same representative. (Redundant numerical representations are difficult to keep consistent, criterion (i).)

(v) *The representation should use features presented in the initial data.* (This is mildly related to Bench-Capon's criteria of 'heuristic adequacy' and of 'notational convenience'.)

(vi) *The representation should allow us to check the desired properties.* Otherwise there should be a translation to another representation using these properties. (This corresponds to Bench-Capon's criteria of 'computational tractability' and 'heuristic adequacy'.)

(vii) *Numerical representations should be continuous.* Small changes in the configuration should result in small changes in the numerical values of features. The computations should be robust, and not prone to numerical errors. (This does not correspond to a general AI criterion.)

Example 5. The Cartesian representation fits criteria (i), (ii), (iv) and (vi). For criterion (vii), the representation is continuous, and generally robust. However, deciding whether three points are collinear etc. may depend on numerical accuracy. For criterion (iv) we selected the unique representative by choosing an origin, axes etc.. Unfortunately, some contexts present essential problems for criterion (v). In addition, criterion (vi) presumes that other representations with alternate features of the configuration are well understood. These are two reasons for examining other representations. ▮

Example 6. The straight-edge constructions fit criteria (i), (iii), (iv), (vii). (For criterion (i), certain degenerate configurations may require auxiliary points - but this is not a serious flaw.) We have seen that constructions may give *incomplete representations*, failing criterion (ii). Of course criterion (v) and (vi) always depend on the context - which data is given and which properties are to be checked. For synthetic geometry, these constructions are adequate, when complete. ▮

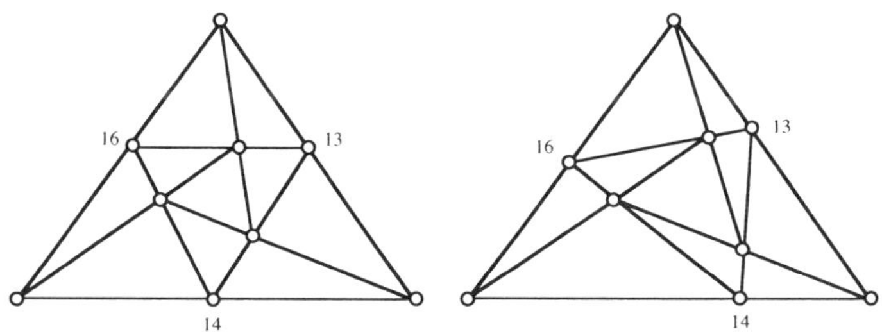

A B

Figure 8.

Example 7. With added incidences beyond a construction, we encountered several problems. The extra incidence of Figure 4A was inconsistent (criterion (i)). The added incidence of Figure 4B was redundant, even for an incomplete representation. The incidences of Figure 5B and 8A do satisfy criteria (i) and (iv), but the result is not projectively unique (Figure 8B). This still violates criterion (ii), and is incomplete. ▮

Incomplete representations are surprisingly common in geometric practice. They represent a stage in the analysis of an object, and raise the problem of listing all possible completions (or equivalently, listing all other data which is already redundant). Of course adequate representation at one level of geometry will become incomplete if we remove some transformations, and become redundant if we add more transformations.

In Example 4, we saw that distances were only 'locally adequate' for criterion (ii). That is, each representative fits a locally unique object, although it may fit a discrete set of objects globally. In a related defect, the representations will only be 'generically adequate' in the sense that this local uniqueness applies to the representatives of 'almost all' objects (including an open dense subset of all objects), but fails for a small (measure zero) subset (see Figure 11D). Example 6D indicated that the representation could be made globally unique - but at a cost of 'local redundance' - and the companion problem of consistency of data (criterion (iv)). Such *local representations* are also very common in geometric practice. Section 5 will examine the example of distances in the plane in some detail.

Representations such as distances and angles will also fail criterion (vi) (see the next section). Translation to the Cartesian representative will be NP-hard. This complexity measures the gap between a representation and a 'construction', which gives an implicit translation to Cartesian coordinates. However these representations are important, because they are forced by criterion (v). Up to questions of representing numbers, the representations will be non-deterministically polynomial: a polynomial algorithm will test whether an object fits a representative.

Finally, 'combinatorial representations', such as incidences, are not continuous (criterion (vii)). For example, collinearity (incidence) is discontinuous near the collinear location. However, combinatorial representations eliminate numerical errors, and may improve tractability (criterion (vi)). For this reason, 'matroidal' (partial incidence) representations and algorithms are very important in geometric analysis by computers [Bokowski & Sturmfels 1989, White 1986].

5. Congruence and the Distance Representation of Points in the Plane.

As suggested in the introduction, basic problems of rigidity of structural frameworks, and of reconstruction of molecules from NMR data, use distances between points as a representation, in the plane and in space. We will extract some crucial features of distance representations for plane points. Our goal is to illustrate the complexity inherent even in these 'natural' measurements.

The distance between points is an invariant feature for congruences but not for larger sets of transformations. The set of pairs of vertices whose distance is measured is given by an *abstract graph* $G = (V, E)$ with a set V of *vertices* and a set E of unordered pairs of vertices or *edges*. The points and chosen distances which are given then form a *geometric graph* $G(\mathbf{p})$:

(i) an abstract graph $G = (V, E)$ and

(ii) a *location map* $\mathbf{p} : V \to \mathbb{R}^2$ which gives plane coordinates $\mathbf{p}(v_i) = \mathbf{p}_i = x_i, y_i)$ for the vertices.

As a visual convention, the measured distances are drawn as line segments joining the two points in a figure.

How many distances should we record? To locate $|V|$ vertices, the Cartesian representation, in standard form, uses $2|V| - 3$ numbers. (Take the $2|V|$ coordinates and remove the coordinates of the first vertex - and one coordinate of the second.) The measurements give the *distance map*:

$$f_G : \mathbb{R}^{2|V|} \to \mathbb{R}^{|E|}$$

which measure the length (squared) of each edge in the geometric graph. For a fixed graph, two congruent sets of points will have the same image under the map, satisfying criterion (ii). If this map from the Cartesian representation is continuous on a large open set of the plane, and 'locally' 1-1 (with some possible degeneracies), then a minimal distance representation will use $2|V| - 3$ features. This is the *dimension of representation* for the configurations, under congruences. (A direct analysis of distance, and local uniqueness gives the same dimension [Laman 1970, Asimow & Roth 1981, Connelly 1990].)

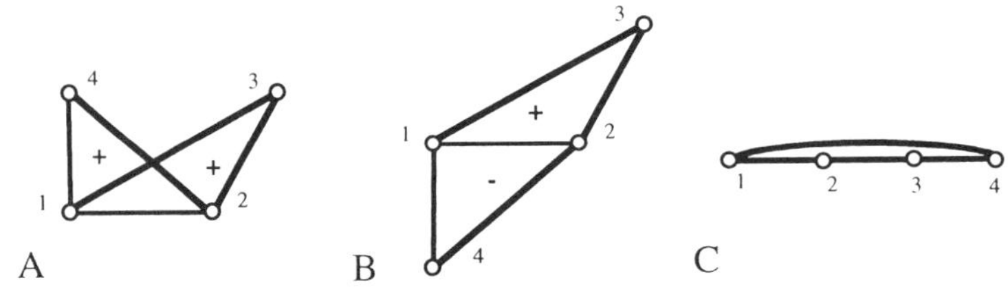

Figure 9.

Will $2|V| - 3$ distances represent a configuration of $|V|$ points?

Example 8. By Example 3, three distances are a good representation for 3 points, while the 5 distances for four points are ambiguous. The two distinct configurations can be separated by an additional sign for the orientation of the triangles (Figure 9A,B). (Unlike Cartesian coordinates, distances are positive numbers - so an additional sign is reasonable.) However, if four points are collinear then the four distances along the line can uniquely represent the points (Figure 9C). ∎

For Cartesian coordinates, any $2|V| - 3$ numbers will represent a geometric set (criterion (i)). However, for distances there are additional inequalities which the numbers must satisfy. The simplest example is the triangle inequality. Writing the distance as $d(a, b)$:

$$d(a, c) \leq d(a, b) + d(b, c)$$

Throughout the rest of this section we will ignore such inequalities. The general problem of deciding whether a set of distances describes a located set of points is known to be NP-hard. For an extensive discussion of these issues, in the plane and higher dimensions, see [Crippen & Havel 1988, chapter 4].

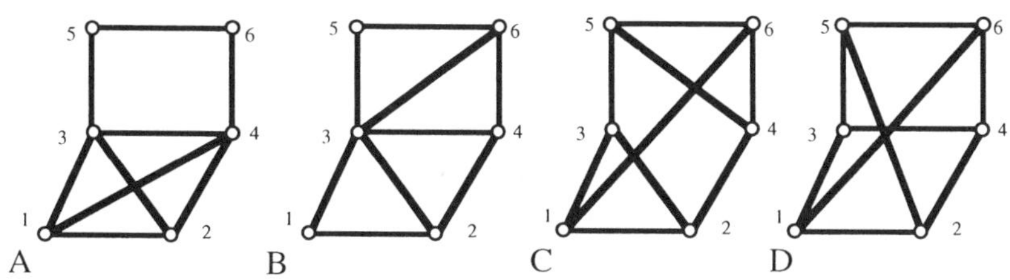

Figure 10.

What about six points? We must select 9 distances among the 15 possible measurements. Some sets are clearly wrong (Figure 10A). Some sets are clearly adequate (with a few sign values to distinguish among the discrete set of locations) (Figure 10B). Two special sets of nine distances are less obvious (Figure 10C,D). The next three examples analyze these patterns.

Example 9. Consider the geometric graphs of Figure 11. These 2-*simple* graphs are created from a single distance, by adding a sequence of 2-valent points (following the order of the vertices). Each graph satisfies $|E| = 2|V| - 3$.

The location of the new point is unique up to reflection through the two attaching points. Thus, for $|V|$ points, the same distances have $2^{|V|-3}$ non-congruent locations. In this case, we can distinguish among the locations by the orientations of these triangles

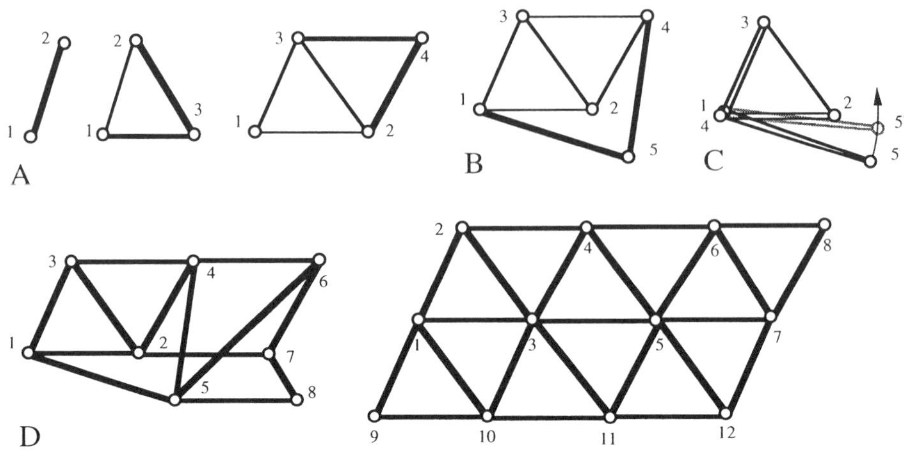

Figure 11.

(relative to the orientation of the original three points). These 2-simple graphs are really 'distance constructions'. We can, up to orientations, find the Cartesian coordinates from the distances, by solving a sequence of two simultaneous quadratic equations [Owen 1991].

Of course there will be degenerate situations (Figure 11C - same graph as Figure 11B), where the two attachments lie at the same spot - and the geometric graph is not well defined, even locally. ∎

The distance map, and its inverse, have been extensively studied as the 'rigidity of plane frameworks' [Asimow & Roth 1978, Connelly 1982, 1990, Whiteley 1983, 1991]. In this abstract structural engineering, the edges are rigid bars, the points are universal joints, and we investigate other locations of the points with the same bar lengths. We translate some basic definitions and results from rigidity into 'distance representations'. Finding the inverse map f_G^{-1} has also been called the 'Fundamental Problem of Distance Geometry' [Crippen & Havel 1988].

If $f_G^{-1}(f_G(\mathbf{p}))$ is a single point in $\mathbb{R}^{2|V|}$, up to congruence, the geometric graph is *globally unique* and the distances $f_G(\mathbf{p})$ represent a unique point set. If $f_G^{-1}(f_G(\mathbf{p}))$ is an isolated point within a neighborhood of $\mathbf{p}$, up to congruence, the geometric graph is *locally unique*, and the distances $f_G(\mathbf{p})$ are a local representation of $\mathbf{p}$. The 2-simple graphs of Example 1, without signs, are locally unique but not globally unique.

If the geometric graph is not locally unique, it is *flexible* (Figure 11D). (Local non-uniqueness actually implies a piecewise analytic path of graphs in $f_G^{-1}(f_G(\mathbf{p}))$ [Asimow & Roth 1978, Connelly 1990].) Flexible geometric graphs are also called 'linkages' or 'mechanisms' in the literature.

The distance map is defined by $|E|$ quadratic polynomials $(\mathbf{p}_j - \mathbf{p}_i)\cdot(\mathbf{p}_j - \mathbf{p}_i) = c_{ij}$. We can study whether the map is a local isomorphism, up to congruence, by studying the Jacobian $\mathrm{d}f_G$ of the map. If this matrix has rank $2|V| - 3$ then the geometric graph is *first-order unique*, and these distances form a good representation of an open neighborhood of the geometric graph [Asimow & Roth 1978, Connelly 1990]. Therefore minimal first-order representations use exactly $|E| = 2|V| - 3$ distances for locations at which the corresponding Jacobian has independent rows. [Other non-linear approaches can prove additional geometric graphs are 'rigid' or locally unique [Connelly & Whiteley 1993] – but these lie outside the scope of this discussion.]

The entries of the Jacobian are algebraic (actually linear) functions of the points and the rank is determined by algebraic formulae (determinants of minors). Therefore there is a single polynomial equation in the coordinates of the points whose zero set gives the geometric graphs where first-order uniqueness fails. (See [White & Whiteley 1984] for a study of these singularities.) In particular, if one geometric graph is first-order unique, then an open dense subset of locations (*almost all*) on the same graph are locally unique. Informally, such locations are *generic* and the graph is a *local representation*. Conversely, if a graph is even locally unique for an open subset of locations, then it is first-order unique on an open dense subset of the locations [Asimow & Roth 1978].

Since these local properties of the map f_G are expressed in a matrix, we sometimes use the vocabulary of matroids (linear algebra) and call a geometric graph with dependent matrix rows a *dependent representation* [Whiteley 1993]. The graph is *dependent* if all locations of a given graph are dependent.

Unfortunately, the same graph and distance values may come from both a locally unique geometric graph, and a flexible geometric graph (Figures 11B and C). Thus a *set of distances* is not 'locally unique', the set of distances is a 'locally unique' representation of a specific location. Of course, a set of distances can be a 'globally unique' representation.

Example 10. Consider the geometric graphs in Figures 12A,B,and C, which locate the same underlying graph, with $|E| = 9 = 2|V| - 3$. The first is globally unique [Connelly 1982], the second is first-order unique [White & Whiteley 1984], and the third is flexible.

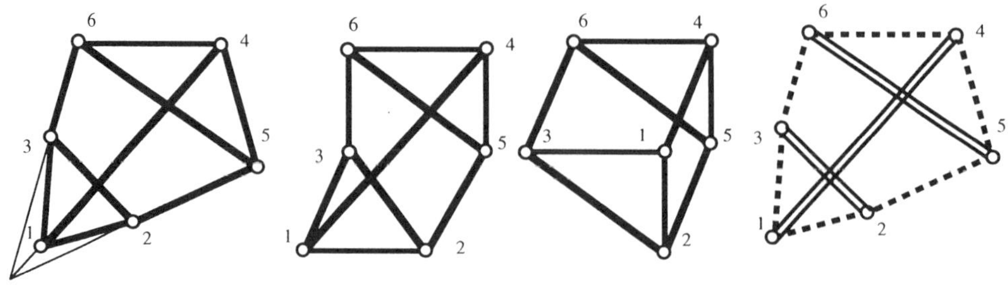

A B C D

Figure 12.

As stated above, the first-order unique locations (such as Figure 12B) form an open dense subset of the entire space of Cartesian realizations $\mathbb{R}^{12}$. The geometric graphs which are not first-order unique either have a collinear triangle $1, 2, 3$ or $4, 5, 6$, or have the three lines $1, 4$, $2, 5$ and $3, 6$ concurrent (projectively) (Figures 12A,C) [White & Whiteley 1984].

There is smaller open dense subset of $\mathbb{R}^{12}$, for which $f_G^{-1}(f_G(\mathbf{p}))$ is a set of isolated points (up to congruence). For all first-order unique locations of the graph, the distance representative is robust - a small change in the values of the distances causes only a small change in the Cartesian coordinates (i.e. linear in the distance changes). There is a local parametrization of the space of locations by the distances.

The globally unique geometric graphs (Figure 12A) form a small (measure zero) closed set. They cannot be first-order unique [Connelly 1982]. Moreover, certain distances are at extreme values - if all other distances are fixed, the last distance can only be decreased (double lines in Figure 12D) or increased (dashed lines). This representative is not robust: some changes in the distances will not be realizable, and for others the resulting deformation in location will not be 'large' compared to the change in lengths. ∎

Example 11. Consider the frameworks in Figures 13A,B,C, which locate a second graph, with $|E| = 2|V| - 3$. This is the bipartite graph $K_{3,3}$, with no triangles. Again, the first is globally unique [Connelly 1982], the second is first-order unique [Whiteley 1983], and the third is flexible.

Once more, the set of first-order unique geometric graphs (such as Figure 13B) forms an open dense subset of $\mathbb{R}^{12}$. The locations which are not first-order unique are now

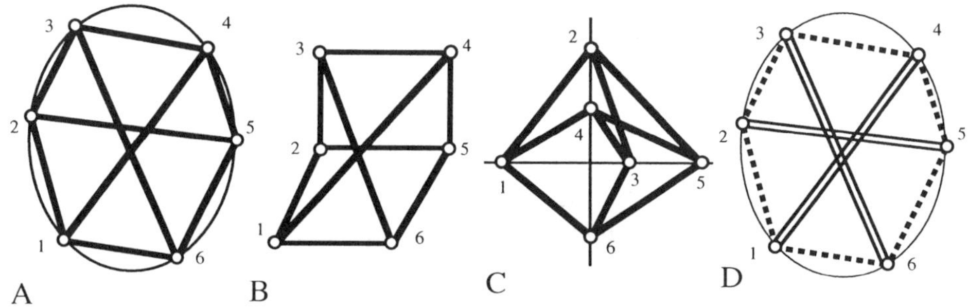

Figure 13.

characterized by a conic section through the six points (including a degenerate section of two lines) [Whiteley 1983].

Again, the globally unique locations are not first-order unique. For a globally unique location, the pattern of extreme values for distances is indicated in Figure 13D.

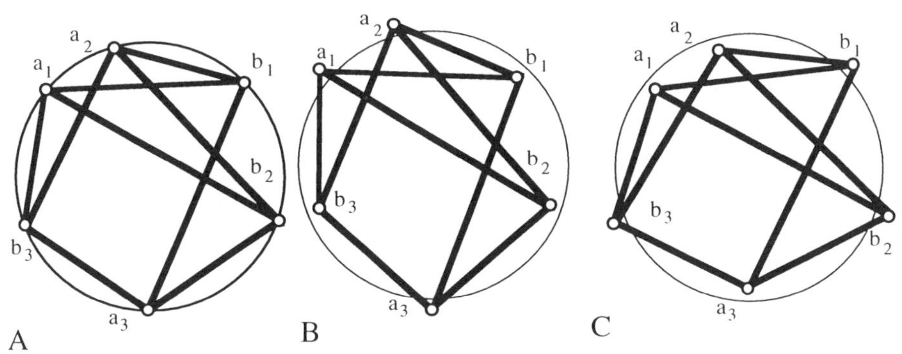

Figure 14.

At locations which are not first-order unique, the map f_G 'folds over' in $\mathbb{R}^{|E|}$ and the local parametrization breaks. Figure 14B,C shows two locations, with identical edge-lengths, related to the location of Figure 14A [Whiteley 1983]. ∎

Remark 2. Two locations with the same set of distances need not be distinguished by different 'signs' or orientations of triangles of points (criterion (ii)). The two locations of Figure 14B,C have the same orientations. Orientations will distinguish the 2-simple geometric graphs, but they do not work in general.

In practice, a plane geometric graph is often represented by an 'engineering drawing'. This is a moderately accurate plane picture - with an explicit set of 'dimensions' (distances, angles etc.), sufficient to select a unique location 'near' the drawing. An accurate knowledge of local representations is required for such drawings, including drawings in computer-aided design. ∎

Remark 3. The 2-simple constructions are, essentially, an algorithm for the (local) inverse map f_G^{-1}. There is no corresponding constructive inverse map f_G^{-1} for the good locations of the graphs of Examples 7 and 8. (Example 7 can be translated into solving a sixth degree equation.)

The (square of the) distance is a polynomial of the Cartesian coordinates. On the other hand, there is no general polynomial algorithm for finding the Cartesian representation from the set of distances and the graph. In principle, this inverse problem has an algorithm (using Tarski's decision procedure for elementary algebra). However, even the following special case is NP-hard:

Given one location for a set of distances, find all other locations for these distances.
See [Crippen & Havel 1988] section 4.4 for a discussion. Even with more than $2|V| - 3$ distances, finding the location is non-trivial. This problem has been studied extensively for the reconstruction of molecules from NMR data. ∎

These examples illustrate basic properties of locally unique, non-redundant distance representations of points by $\mathbb{R}^{2|V|-3}$ distances. First, these distances must be 'spread out' in the configuration (recall Figure 10A). Specifically, in the graph of distances:

For all non-empty subsets E' of edges, incident with vertices V':

$$|E'| \leq 2|V'| - 3. \tag{1}$$

Second, while 'almost all' locations of the graph are robust local representatives, these are not globally unique. The globally unique locations will represent a failure of the local parameters.

The necessary condition of equation (1) is also sufficient [Laman 1970, Asimow & Roth 1978].

Laman's Theorem 4. *For any graph $G = (V, E)$ satisfying:*

(i) $|E| = 2|V| - 3$;

(ii) For all non-empty subsets E' of edges, incident with vertices V':

$$|E'| \leq 2|V'| - 3.$$

almost all geometric graphs $G(\mathbf{p})$ are first-order unique.

A graph with these properties is a *congruence graph*. Figure 15 shows some other examples of congruence graphs.

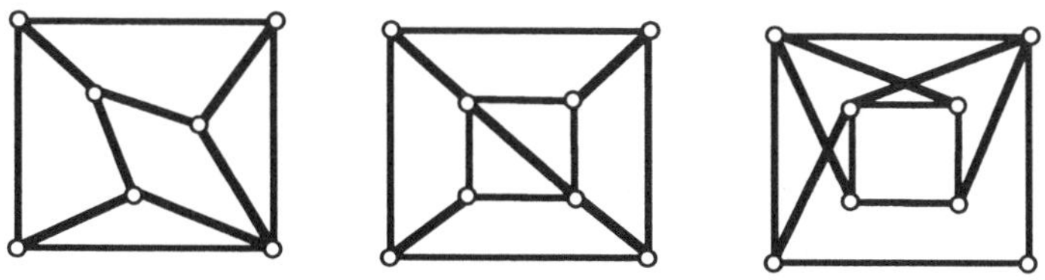

Figure 15.

Remark 5. The condition of Laman's Theorem, running over all subsets E', appears to be an exponential condition. However there are several standard polynomial algorithms (order $O(|V|^2)$) which check the condition [Sugihara 1985]. Of course this algorithm is combinatorial, not numerical. The singular locations in the plane which are not locally unique are not described by this algorithm [White & Whiteley 1984].

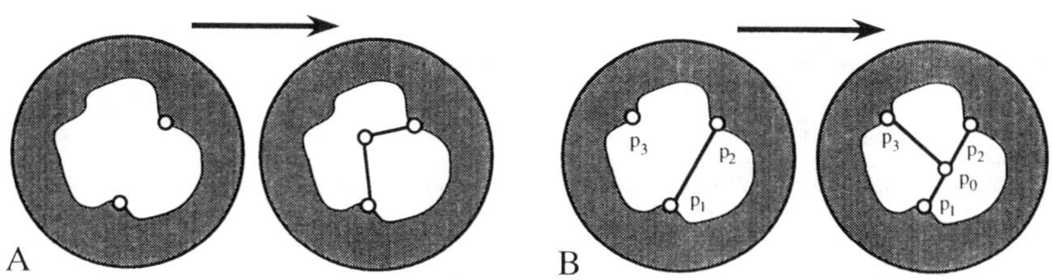

Figure 16.

All the congruence graphs can be constructed from a single edge, by an inductive sequence using two steps (Figure 16). (The first step alone constructs the 2-simple graphs.) These are the *Henneberg 2-sequences* presented in [Tay & Whiteley 1985]. ▮

Some graphs will give globally unique, but redundant, representations with $2|V|-2$ distances (Figure 17), for almost all locations. Conjectured conditions for this are [Hendrickson 1990]:

(a) *the graph is 3-connected (to prevent reflections of one portion);*

(b) *for all proper subsets V' of vertices, containing edges E',*

$$|E'| \leq 2|V'| - 3.$$

▮

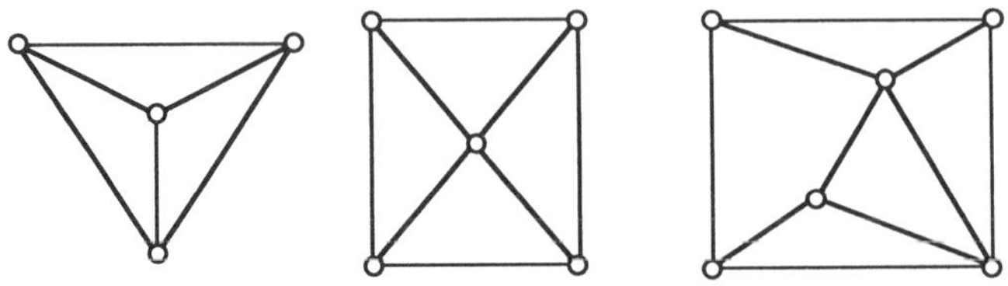

|E|=2|F|-2, |E'| ≤ 2|F'|-3

A B C

Figure 17.

Finally, the property, that a geometric graph and a set of distances is a locally unique representation, is projectively invariant [Roth & Whiteley 1981]. That is, if two geometric graphs are projectively equivalent then the corresponding set of distances are either both first-order unique or both fail first-order uniqueness. This 'projective invariance' is not true for global uniqueness.

Remark 6. There is a well developed theory for distance inequalities as constraints on a plane geometric graph - to obtain uniqueness or local uniqueness. In rigidity, these are called 'tensegrity frameworks' [Roth & Whiteley 1981]. We will not describe the details - except to not that such extensions have bee studied. ∎

We have tried to illustrate some of the complexities of distance representations for points in the plane. A non-redundant distance representation is locally unique, and robust, but there are degeneracies. The graphs which will serve as representations have a combinatorial characterization, but the degenerate locations require exponential calculations. The inverse map, from distances to locations, is NP-hard. The primary reason to use these distance representations is that they arise naturally from our tools of measurement.

As the next sections show, other common representations are even more difficult to analyze. In comparison, the theory of plane distances provides a model of effective information, which other geometric theories seek to emulate.

6. Angles and Similarities for Plane Sets.

A surveyor will draw a map, up to choice of scale, from a set of lines and angles. Specifically, the angle $\theta(a, b; c, d)$ between two lines ab and cd is invariant under similarities. For criterion (vii) (continuity), we assume that the domain for each angle is the real numbers modulo 2π. Since the dilations were added to the congruences, the dimension of representation, up to similarity, is now $2|V| - 4$. What sets of angles are an adequate representation for a geometric graph?

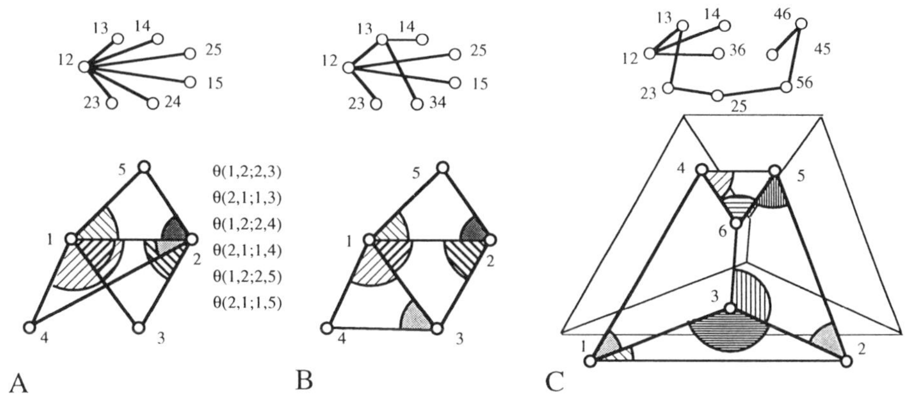

Figure 18.

Example 12. Building up triangles creates some correct angle sets (Figure 18A and B). These 'similarity patterns' correspond to the 2-simple geometric graphs. The angles have a sign, recording the orientation, so that these simple representations are globally unique, not just locally unique. This 'triangulation' construction used in surveying matches the count $|\Theta| = 2|V| - 4$. ∎

There is an underlying 'tree property' for indepndent angles. An angle connects two edges of the original geometric graph. With the original edges as 'vertices' of a graph, the angles become 'edges' of the *angle graph* (see Figure 18). Any *cycle of angles*: $\theta(a, b; c, d)$, $\theta(c, d; e, f)$ ... $\theta(j, k; l, m)$, $\theta(l, m; a, b)$ is is redundant. Therefore, the angle graph of a minimal representation (criterion (iii)) must be a forest (free of cycles).

This angle graph has a natural geometric interpretation. Each line has a direction (a point at infinity or on the unit circle). An angle acts like a 'distance' between the two directions. On a line (or on the circle) any polygon of 'distances' is dependent.

A spanning tree of $|E| - 1$ angles in the angle graph fixes directions for the edges of a geometric graph, up to a global rotation. Does fixing the angles also fix the original points locally (or globally) up to similarity? This problem has been studied, implicitly, within the field of 'parallel redrawings' [Whiteley 1988]. Figure 18C shows a non-trivial 'parallel drawing': a pair of graphs in which corresponding edges are parallel and all angles are equal, but the two configurations are not similar. We conclude that it is not enough, in general, to fix the angles between the given edges, in order to obtain uniqueness, up to similarity

If the points are not fixed by these directions, at first-order, then there is a continuous (analytic) path of dissimilar locations with the same directions for the $|E|$ edges. Finally, if any two dissimilar locations have the same directions, then there is a continuous (analytic) path of locations with the same directions for the E edges (non-trivial *parallel redrawings*) (Figure 18C). Local and global uniqueness, up to similarity, coincide for a spanning trees of angles. An unexpected theorem of discrete geometry [Whiteley 1988], shows that, for a spanning tree of angles, the global uniqueness is geometrically equivalent to the first-order uniqueness of the same geometric graph for distances! This gives the result:

Theorem 7. *For any graph $G = (V, E)$ satisfying:*

(i) $|E| = 2|V| - 3$;

(ii) For all non-empty subsets E' of edges, incident with vertices V':

$$|E'| \leq 2|V'| - 3.$$

(iii) there is a spanning tree of $|E| - 1$ angles

then almost all geometric graphs $G(\mathbf{p})$ on these edges are globally unique for the given angles, up to similarity.

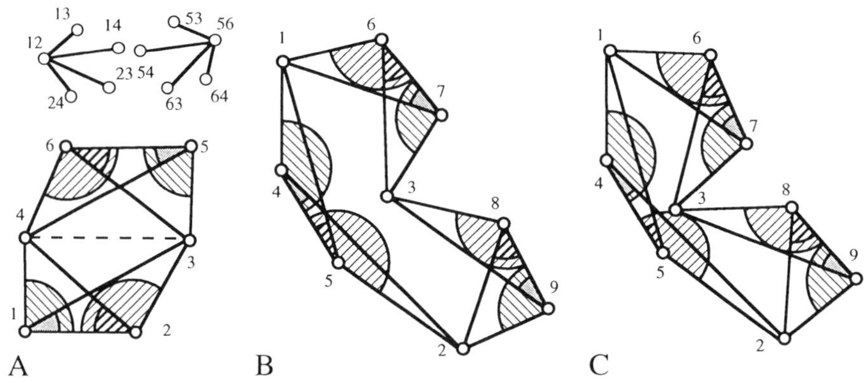

Figure 19.

Of course not all local representations by angles involve a spanning tree. The pattern of Figure 19A uses $|E| - 2$ angles - a forest of two trees on the edges drawn. Moreover, with forests of more trees in the angle diagram, local and global uniqueness are different. The pattern of Figure 19B is locally unique, but the angles match Figure 19C - so global uniqueness fails.

To construct Figures 19B,C, we replaced the edges of a triangle with locally unique angle configurations. If a similar replacement is applied to the geometric graphs of Figure 14B,C, even the orientations of all triangles of points does not distinguish dissimilar patterns with the same angles. Geometrically, angles and similarity reproduce all the difficulties of distance patterns for congruence.

For a general set Θ of angles, a minimal local representative must also satisfy the second condition:

$|\Theta| = 2|V| - 4$ *and for all non-empty subsets* Θ' *of ratios, incident with vertices* V'*:* $|\Theta'| \leq 2|V'| - 4$.

This condition, plus the tree condition, does not characterize the angle representations, as the following example illustrates.

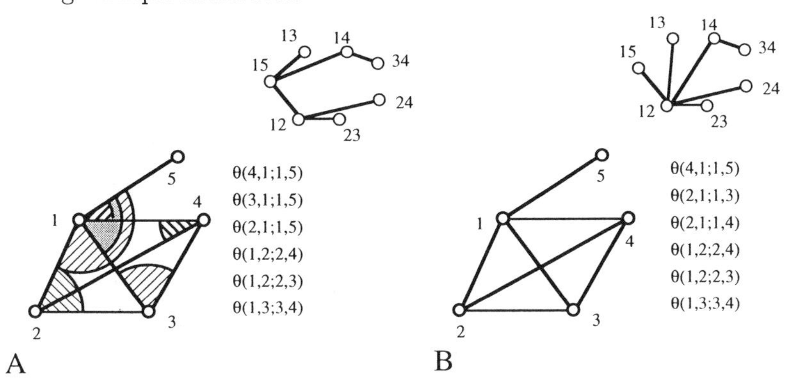

Figure 20.

Example 13. Consider the angles given in Figure 20A. This clearly satisfies the combinatorial conditions - and is not a locally unique representation. It cannot hold the point 5 relative to the other points.

In the angle diagram, any tree is equivalent to another tree on the same vertices. In this example, such a replacement leads to a new set which violates the original counts (Figure 20B). ∎

An appropriate combination of the counts and the tree replacement process is missing.

Remark 8. The results on distances, which are 'variant' for similarity maps, can also be adapted to similarity by using 'ratios of distances' (which are invariant for similarity).

Example 14. Consider any congruence graph G, and a locally unique geometric graph $G(\mathbf{p})$. If one distance divides all of the others, the ratios are locally unique, for similarity. (Fix the first distance - and the object is fixed up to congruence. The fixed distance just removed the dilations.) These patterns correspond to the angle patterns of Theorem 6.

A sequence of divisions on the congruence graph, beginning with one distance, and choosing the next denominator from the set of previously used distances will also work.

Each ratio pairs two edges of the original graph. With the original edges as 'vertices' of a graph, the ratios become 'edges' of a *ratio graph*. The construction of Example 11 creates a spanning tree in the ratio graph. Figure 21 shows some ratios and ratio graphs. (The two ratios $r(a,b;c,d)$ and $r(c,d;a,b)$ are equivalent, so this is an undirected graph.)

The minimal *ratio sets* Ra representations must satisfy the same combinatorial conditions as angles:

(i) the ratio graph must be cycle free (a forest).

(ii) $|Ra| = 2|V| - 4$ *and for all non-empty subsets* Ra' *of ratios, incident with vertices* V'*:* $|Ra'| \leq 2|V'| - 4$.

In fact, all combinatorial results for angles are interchangeable with results for ratios of distances. Recent unpublished work on plane CAD demonstrates that:

Theoren 9. *For a fixed vertex set V, a distance ratio set Ra is a local representation of a plane configuration of points for almost all location maps if and only if the corresponding angle set Θ is a local representation of a plane configuration of points for almost all location maps.*

Moreover, we could mix data with some distance ratios and some angles to represent a configuration up to similarity. For example, the ratio and the angle of a pair of adjacent edges $(a,b;b,c)$, fixes the triangle, up to similarity. This raises even more unsolved problems. ∎

Remark 10. For plane geometric graphs, represented up to congruence (or even up to translation), we can mix lengths of edges and directions for edges (a simpler structure than adding angles to distances). This special case has recently been solved [Servatius & Whiteley 1995, Whiteley 1995]. However, the general problem of using distances and angles to represent a plane configuration remains unsolved.

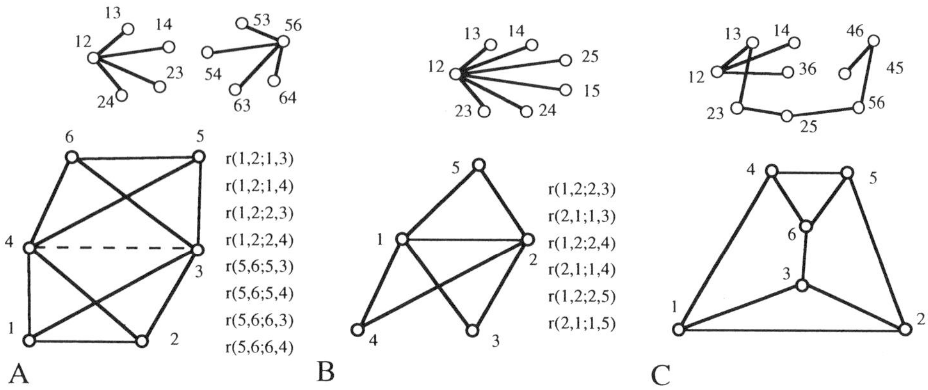

Figure 21.

Nevertheless, such problems do have practical applications. In geodesy (map making), a surveyor's measurements of angles and distances are used to correct a given map. Before launching a long, least squares numerical calculation, it would help to know whether the pattern is locally unique (at least generically). Thus necessary combinatorial conditions are implemented in the algorithms [Snay & Spriggs 1982], and sufficient conditions would be desirable. ∎

In the absence of such an analysis, some computer-aided design programs use 'limited constructions' to generate the 'allowed representations'. This restricts the specifications which a designer can directly choose, for the convenience of the computer programmer. We want to do better and this will require more geometry!

7. Projective Maps and Incidences in the Plane

For projective transformations, we return to the point/line incidences of Section 3. A line not through the origin, can be represented by an equation:

$$ax + by = 1.$$

The line l^j has *Cartesian line coordinates* $P^j = (a^j, b^j)$. A *configuration of points and lines* is an *incidence structure* $S = (V, L; I)$ composed of abstract vertices V, abstract lines L and incidences $I \subset V \times L$, together with a pair of location maps $\mathbf{p} : V \to \mathbb{R}^2$ and $\mathbf{P} : L \to \mathbb{R}^2$, such that, for each $(v_i, l^j) \in I$:

$$\mathbf{P}^j \cdot \mathbf{p}_i = a^j x_i + a^j y_i = 1.$$

This gives $2|V| + 2|L|$ coordinates and I incidences. Since the projective transformations are a space of dimension 8, the set of incidences in a locally unique, minimal representation must have size $|I| = 2|V| + 2|L| - 8$. A natural necessary condition is that every subset of incidences must satisfy: $|I'| \leq 2|V'| + 2|L'| - 8$. If these incidences are independent, they will be a combinatorial (non-numerical) representation.

Example 15. The straight-edge construction of Figure 3A is such a representation, with $|I| = 18 = 2|V| + 2|L| - 8$. The extended structure of Figure 4A, with $|I| = 21 > 20 = 2|V| + 2|L| - 8$, is redundant. The construction of Figure 3B has $|I| = 24 < 26 = 2|V| + 2|L| - 8$, so it is not a complete representation.

The extended incidence structure of Figure 5B still has $|I| < 2|V| + 2|L| - 8$, so it is a partial representation. ∎

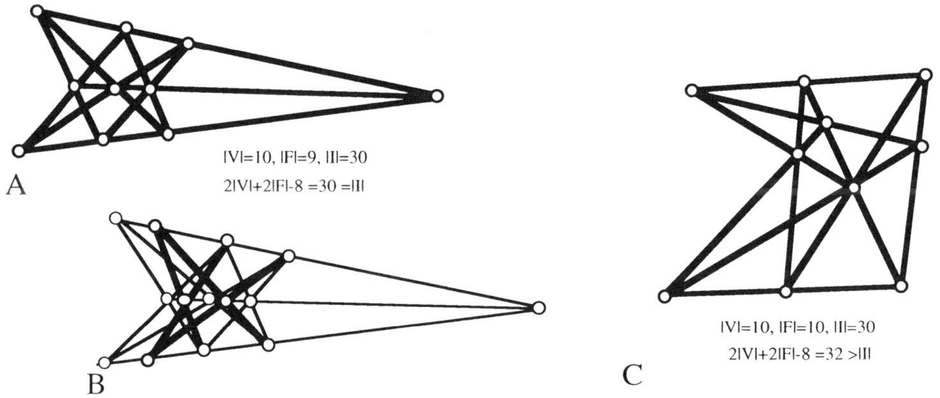

Figure 22.

Example 16. Consider the configuration of Figure 22A, with $|I| = 2|V| + 2|L| - 8$. This contains, as a substructure, the Pappus' configuration of Figure 4B. The incidences are dependent, and the configuration is not fixed, up to projection. Figure 22B indicates a path of configurations with the same incidences.

Figure 22C shows the basic diagram for Desargues' Theorem, with $|I| < 2|V| + 2|L| - 8$. Since the last incidence is guaranteed by Desargues' Theorem, the incidence structure is also dependent. ∎

However, the problems of incidence structures are even more extreme.

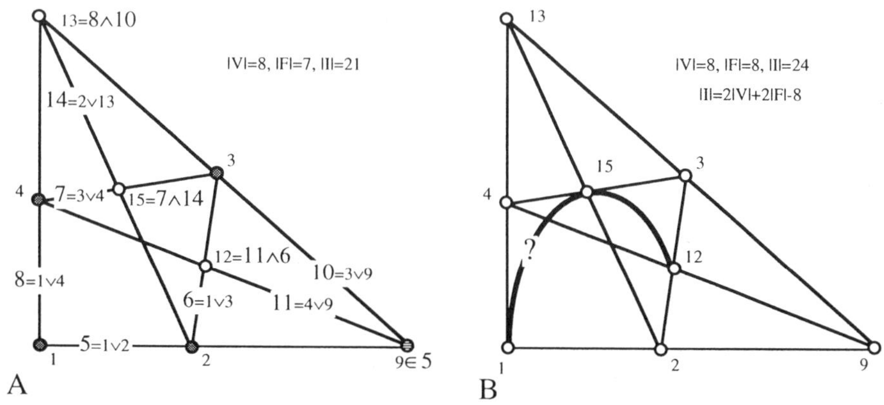

Figure 23.

Example 17. The configuration of Figure 23A is a construction with $|I| < 2|V|+2|L|-8$. This is independent, but an incomplete representation. The extra choice, beyond the first four points, is the point 9 on the line 12.

With one more line and three more incidences (Figure 23B) the structure satisfies $|I| = 2|V| + 2|L| - 8$. However, the resulting incidence structure can only be located in the complex plane. This does *not* a represent a real configuration! ∎

It seems improbable that we could characterize all representing sets of incidences in this general setting. Each theorem of projective geometry is the claim that a final incidence follows from previous incidences. All the complexities of algebraic equations appear within projective configurations. There are also configurations with a discrete set of distinct locations, indistinguishable by the orientations of all triples of points.

Conjecture 11. *There is no polynomial-time decision algorithm for which incidences structure are independent for some (almost all) realizations as points and lines in the plane.*

Remark 12. It is possible to combine general incidences (as an incomplete representation from a construction) with additional distances, angles or both. This is the natural setting for constrained or parametric CAD. In fact some programs include a feature that identifies (by colour) additional constraints which would complete the representation to a locally unique diagram, as well as blocks of information which are now complete (i.e. in which any additional constraint would be redundant).

With one exception, we have not specific results on which data can be added to an incomplete incidence representation to complete it. The one exception is, in fact, a plane version of general theorem in scene analysis [Whiteley 1988].

The plane configurations is represented by giving the incidence structure and the (central) projection onto a fixed line (say $y = 0$) - a *line picture*.

Line Picture Theorem 13. *An incidence structure $S = (V, L; I)$ with the vertical projection x_i for all vertices, is a generically correct representation, up to vertical scale and vertical translation, if and only if $|I| = |V| + 2|L| - 3$ and for all subsets $|I'|$ of incidences for at least two vertices, $|I'| \leq |V'| + 2|L'| - 3$.*

Unlike the subtler geometry for distances, these results generalize immedateley to plane pictures of spatial configurations, and similar ideas in all higher dimensions.

With more incidences and projections, or with distances, angles, and incidences, there are essential unsolved problems. If these alternate patterns are to be used in computer geometry, a great deal of plane geometry remains to be done. ∎

Remark 14. Any 'good' representation for points and lines in the projective plane should finally be invariant under polarity: a polarity, switching points and lines, should take one representation to a new representation of the polar configuration. This holds for a representation by homogeneous coordinates (for lines and points) and for a representation by an incidence structure.

In this spirit, the representation of points and lines by projections of points should be paired with a representation of points and lines by the sections of lines with a fixed line. While initial results for representation by sections do exist, the general problem of mixed sections and projections has not been thoroughly investigated. ∎

8. Conclusions.

Geometers have been 'representing' objects for a long time. There are many other geometric studies which can be translated into questions of local or global representations of geometric configurations [Whiteley 1994]. Here I have selected a few plane representations closest to my own experiences.

In closing, we compare the criteria of Section 4 with the geometric representations of Sections 5-7.

(i) *Each representative should fit some object (it should not contradict the axioms of the geometry).*

For general distances, angles etc., we did not characterize the numerical values for distances which would be appropriate. The cases where this criterion was met were: Cartesian representations; simple geometric graphs (using the triangle inequality); and a tree of angles and incidences.

(ii) *A representative should fit a 'unique object'.*

This criterion was modified to 'local uniqueness' - and even 'generic local uniqueness' for distances, angles, etc.. First-order uniqueness was used, because it was tractable and because it implies local uniqueness for all locations in a neighborhood. Global uniqueness occurred for the Cartesian representation and for a spanning trees of angles with incidences.

This criteria is also violated by incomplete representations. For projective constructions, the essential problem is choosing appropriate completions.

(iii) *A class of equivalent objects should have a single representative.*

This was built into the representations used (assuming the ordering, or labelling of the points, lines, planes, etc.). Beyond the Cartesian coordinates, we chose invariant features.

(iv) *The representative should be minimal (non-redundant).*

This was achieved, at least to first-order, by the combinatorial counts of the 'dimension of local representation'. Global uniqueness, without redundance, is typically an unsolved problem. Other examples, such plane pictures, are usually redundant, raising the computational problem of extracting an 'independent' representation for computational consistency.

(v) *The representation should use features available in the initial information which is presented or measured.*

This was an essential feature of the representations we presented. The selected features, such as distances, angles etc., originated in particular applications and their corresponding data.

(vi) *The representation should allow us to check the desired properties.*

This criterion is met for Cartesian coordinates and for plane pictures (using linear algebra). Also for representations based on 'constructions', such as simple geometric graphs, and straight-edge constructions, the Cartesian coordinates are found by a polynomial time from the representation. Beyond this, the inverse problem becomes NP-hard.

(vii) *The representation should be continuous.*

This was provided by all of the numerical representations, passing from the Cartesian to another representation. Unfortunately, there is no globally defined, continuous inverse passing from other representations, such as distances, etc. back to Cartesian coordinates.

In summary, criterion (v) (features derived from the application) is in conflict with criteria (ii) (a unique object for each representative) and (vi) (translations to other representations). Resolving this conflict poses fundamental geometric problems.

All of these difficulties, and more, arise for representations of spatial configurations. Whiteley 1994 provides a related summary of results, conjectures and unsolved problems, for the special class of spherical polyhedra in 3-space.

We hope that the examples indicate the traps which await those who niavely select 'geometric features' as a 'representation' of a class of geometric objects. Most of all, we have illustrated some fundamental unsolved problems in geometry, problems worthy of further investigation.

Acknowledgments. Vincent Mirelli posed some initial questions at the workshop on Learning and Geometry, and spent a stimulating evening discussing my preliminary responses.

I learned this geometry with Janos Baracs and Henry Crapo at the Structural Topology Research Group, Université de Montréal. Over the last 15 years, we worked together on the rigidity of frameworks, on the construction of spatial polyhedra, on pictures of spatial polyhedra, and on determination of polyhedral designs. Recent work on plane CAD with John Owen and Brigitte Servatius has added another perspective on the underlying problems. From these experiences, I have been repeartedly reminded that selecting a 'good representation' of geometric objects is a central problem of applied geometry.

Bibliography

[Asimow & Roth 1978]

L. Asimow and B. Roth, Rigidity of graphs, *Trans. Amer. Math. Soc.* 245, 279-289.

[Baracs 1988]

J. Baracs, Spatial perception and creativity; in **Shaping Space: A Polyhedral Approach**, (M. Senechal and G. Fleck eds.), Birkhauser, Boston, 118-132.

[Bench-Capon 1990]

T. J. M. Bench-Capon, **Knowledge Representation: An approach to Artificial Intelligence**, The APIC Series 32, Academic Press, London, San Diego.

[Bokowski & Sturmfels 1990]

J. Bokowski and B. Sturmfels, **Computational Synthetic Geometry**, Lecture Notes in Mathematics 1355, Springer-Verlag, Berlin, Heidelberg, New York.

[Chou 1988]

S-C. Chou, **Mechanical Geometry Theorem Proving**, Mathematics and its applications Vol 41, D. Reidel Pub. Co. ; Norwell, MA, U.S.A.

[Connelly 1990]

R. Connelly, Fundamental concepts of rigidity, to appear in **Geometry of Rigid Frameworks**, H. Crapo and W. Whiteley (eds).

[Connelly & Whiteley 1993]

R. Connelly and W. Whiteley, Second-order rigidity and pre-stress stability for tensegrity frameworks, *SIAM J. Discrete Mathematics*, to appear.

[Crippen & Havel 1988]

G.M. Crippen & T.F. Havel, **Distance Geometry and Molecular Conformation**, Research Studies Press, Somerset, England (Distributor John Wiley and Sons, New York).

[Forsyth et al 1990]

D.A. Forsyth, J.L. Mundy, A.P. Zisserman and C.M. Brown, Invariances - a new framework for vision, internal report, Oxford University Engineering Laboratory.

[Graver et al 1993]

J. Graver, B. Servatius and H. Servatius, *Combinatorial Rigidity*. AMS Monograph 2, 1993.

[Hendrikson 1990]

B. Hendrikson, The Molecule Problem: Determining Conformation from Pairwise Distances, Ph.D. Thesis, Department of Computer Science, Cornell University, Ithaca, New York 14853.

[Hilbert & Cohn-Vossen 1956]

D. Hilbert and S. Cohn-Vossen, **Geometry and the Imagination**, Chelsea, New York. Translation of **Anschauliche Geometrie**, Springer, Berlin 1932.

[Klein 1872]

F. Klein, **Vergleichende Betrachtungen über neure geometrische**, Forschungen, Erlangen, (*Math. Annalen* 43, p63-).

[Klein 1948]

F. Klein, **Elementary Mathematics from an Advanced Standpoint: Geometry**, English translation, Dover, New York.

[Laman 1970]

G. Laman, On graphs and the rigidity of plane skeletal structures, *J. Engineering Math.* 4, 331-340.

[Owen 1991]

J.C. Owen, Algebraic solutions for geometry from dimensional constraints, Symposium on Solid Modeling Foundations and CAD/CAM applications, ACM press, 1991

[Owen 1993]

J.C. OwenConstraints on simple geometry in two and three dimensionspreprint, D-Cubed Ltd., 68 Castle Street, Cambridge CB3 OAJ, UK

[Roth & Whiteley 1981]

B. Roth and W. Whiteley, Tensegrity frameworks, *Trans. Amer. Math. Soc.* 265, 419-445.

[Servatius & Whiteley 1995]

B. Servatius and W. Whiteley, Constraining Plane Configurations in CAD: combinatorics of directions and lengths. Preprint, York University, North York, Ontario, 1995.

[Snay & Spriggs 1982]

R. Snay and J. Spriggs, An algorithm for testing the solvability of horizontal networks, preprint Nat. Geod. Survey, NOAA Rockville Md. 20852.

[Sugihara 1985]

______, Detection of structural inconsistency in systems of equations with degrees of freedom, *J. Disc Appl. Math.* 10, 297-312.

[Tay & Whiteley 1985]

T-S. Tay and W. Whiteley, Generating all isostatic frameworks, *Structural Topology* 11, 21-69.

[White 1986]

N. White (ed), **Theory of Matroids**, Encyclopedia of Mathematics and its Applications 26, Cambridge University Press, Cambridge, London, New York.

[White & Whiteley 1984]

N. White and W. Whiteley, The algebraic geometry of stresses in frameworks, *SIAM Journal of Algebraic and Discrete Methods* 4, 481-511.

[Whiteley 1983]

W. Whiteley, Infinitesimal motions of a bipartite framework, *Pac. J. of Math.* 110, 233-255.

[Whiteley 1988]

______, Some matroids on hypergraphs with applications to scene analysis and geometry, *Disc. and Comp. Geometry* 4, 75-95.

[Whiteley 1991]

W. Whiteley, Logic and invariant computation for analytic geometry, *J. Symbolic Computation* 11, Special Issue on Invariant Theoretic Methods in Symbolic Computational Geometry, 549-578.

[Whiteley 1993]

______, Matroids and rigidity, in **Matroid Applications**, Neil White (ed.), Cambridge University Press, pages 1-53.

[Whiteley 1994]

______, How to describe or design a polyhedron. *J. Intelligent & Robotic Systems* 11,135-160.

[Whiteley 1995a]

______, Synthetic vs analytic geometry, this volume.

[Whiteley 1995b]

______, Constraining Plane Configurations in CAD: geometry of directions and lengths. Preprint, York University, North York, Ontario.

[Wu 1984]

Wu Wen-tsün, Basic principles of mechanical theorem-proving in elementary geometries, *J. of Sys. Scie. & Math. Scies.* 4, 207-235.

Geometry Theorem Proving in Euclidean, Descartesian, Hilbertian and Computerwise Fashion

*Wu Wen-Tsün**

(Institute of Systems Science Academia Sinica)

January 7, 1992

*The present paper is the completed version of a talk given in Learning and Geometry Workshop, Jan. 8-11, 1991, organized by University of Maryland at College Park (UMCP). The author would like to express his warm gratitude to Systems Research Center of UMCP in inviting him for one-month stay and in offering him an agreeable environment to complete his work. The author would like also to express his warm thanks to Ms Cheri Helms for her nice typing of the paper

The evolution and development of geometry theorem-proving, dating from Euclid's "Elements" in 3c B.C., may be divided into several stages as indicated in the title of the paper. Achievements in recent years due to the Mathematics-Mechanization Group of the Institute of Systems Science, Academia Sinica are also briefly described.

KEY WORDS. Mechanical theorem proving. Mathematics mechanization. Geometry theorem proving. Mechanizability. Mechanization theorem.

1. Introduction

In Euclidean geometry there are innumerous beautiful theorems bearing such eminent names like Gauss, Desargues, Pappus, Pascal, Feuerbach, Morley, et al, to mention just a few of them. To prove such theorems offers an immensely delightful challenge to the mental power of a human. However, in tracing the historical tracts of such theorem proving one may see that there are several stages of developments in dealing with such mental activities in various different fashions which have evolved since the understanding of the essence of the activities deepens as time goes on. To see how in different stages the same theorem will be proved in different ways let us consider a particular theorem as shown below.

The theorem in question is due to K.F. Gauss. By a complete quadrilateral will be meant a geometrical configuration consisting of 4 lines in the plane meeting each other in 6 different points. Cf. the Fig 1 in which the 4 lines forming the complete quadrilateral are AEB, AFD, BCF, and CDE. The three pairs of the intersecting points $(A, C), (B, D)$, and (E, F), not lying both on any such lines are then called diagonal pairs. Gauss, in a short paper [G], proved that the centers of all conics touching the 4 lines of such a quadrilateral are co-linear. As the 3 diagonal pairs may be considered each as a degenerate line conic touching these 4 lines, Guass deduced as a corollary the following.

Theorem G. *The mid-points of 3 diagonal pairs of the complete quadrilateral are co-linear.*

The first and earliest stage of geomerty theorem-proving is the Euclidean fashion, which is familiar to everyone since one's youth as a school child. A proof of theorem G in such fashion may be given as follows. Through the points B, C, D, E, F let us draw parallels to the lines AEB and AFD to form parallelograms $AB\prime CD\prime, ABH\prime D$, and $AEI\prime F$ as shown in Fig. 2. From known theorems about proportionality properties induced by parallel lines one proves first that the points, $C, H\prime, I\prime$ are co-linear. The points G, H and I being the mid-points of the diagonals AC, BD, and EF of the above parallelograms, are also mid-points of the diagonals $AC, AH\prime$, and $AI\prime$ of these parallelograms. Since $C, H\prime, I\prime$, are co-linear, so are the mid-points G, H, I, as was to be proved.

From the above we see that proofs of this fashion rely on the Definition-Axiom-Theorem-Proof deductive system as described in Euclid's classic "Elements" of 3 c B.C. Each step in the proof is a logical consequence of facts already known either as an axiom or a theorem previously proved in the same manner. In the labyrinth of possible steps to follow one should seek a way to lead from the hypothesis of the theorem step by step up to the conclusion of the theorem. Usually auxiliary lines should be drawn to facilitate reasoning. There are no general rules for the construction of such auxiliary lines, nor are there any for the steps of reasoning to follow. This round-about way of proving theorems requires the highest ingenuity and cleverness as well as the indescribable talents of the geometer. Moreover, the proof of one theorem can hardly be applied to the proof of any other theorem, even though they may be similar to each other in appearance. In fact, thousands of theorems require thousands of proofs as well as constructions of auxiliary lines.

These are the characteristic features of the first stage geometry theorem proving in the Euclidean fashion, which we shall reformulate as (a) and (b) below:

(a) Proving is carried out in a deductive way by logical reasonings based on a Definition-Axiom-Theorem-Proof System.

(b) Proving is carried out for theorems one by one, i.e., each individual theorem has an individual proof peculiar to the theorem itself.

2. Stage of Geometry Theorem-Proving in Descartesion Fashion

The appearance of the classic "La Géomètrie" by Descartes in 1637 made possible an utterly new fashion of geometry theorem-proving. In the French edition of the classic Descartes wrote at the very beginning on the margin "Comment le calcul d'Arithmétique se rapporte aux opérations de Géomètrie". He explained in the text how geometrical constructions may be carried out corresponding to the main arithmetical operations $a + b, a - b, ab, a/b$, $\sqrt{a^2 + b^2}$, etc. for segments of lengths a, b. He then explained, with an implicit coordinate system, how the method can be applied to various problems arising from geometry and optics. The creation of this now-a-days so-called analytical geometry has thus paved the way for reducing geometry to algebra or arithmetic in Descartes's own terminology and replacing logical reasonings by mere algebraic computations. For a geometry theorem one may formulate, by the introduction of coordinate systems the hypothesis and conclusions of the theorem as sets of equations of the form

$$f_i(x_1, \cdots, x_n) \begin{matrix} \geq \\ = \\ \leq \end{matrix} 0 \text{ and } g_j(x_1, \cdots, x_n) \begin{matrix} \geq \\ = \\ \leq \end{matrix} 0$$

resepectively (to be abbreviated *resp.* in what follows). The proof of the theorem consists of then seeking the way of deducing the equations for g_j from those for f_i by algebraic manipulations of the latter ones. Geometry theorem-proving is thus reduced to some computational work. To see how this is to be done let us consider again the theorem G for illustration. It seems that Gauss was very proud of his theorem G, and he considered it to be entirely new. He wrote:

> *Diese allgemeine Eigenschaft eines jeden Vierecks ist meines Wissens nicht bemerkt, ich werde davon unten einen enfachen direkten Beweis geben.*

We reproduce now Gauss' own proof below, using Gauss', own notation. Let us take an arbritrary coordinate system, and let the various points be $A = (a, a\prime), B = (b, b\prime)$, etc. Similary let the mid-points be $G = (g, g\prime), H = (h, h\prime)$, and $I = (i, i\prime)$ so that

$$g = \frac{1}{2}(a + c), g\prime = \frac{1}{2}(a\prime + c\prime), \text{etc}$$

The geometrical condition that A, E, B should be co-linear will be turned to an algebraic equation, viz.

$$a(e\prime - b\prime) + b(a\prime - e\prime) + e(b\prime - a\prime) = 0$$

Similarly the conditions that A, F, D *resp.* B, C, F *resp.* C, E, D are co-linear will be turned *resp.* into algebraic equations

$$a(f' - d') + d(a' - f') + f(d' - a') = 0$$

$$b(c' - f') + c(f' - b') + f(b' - c') = 0$$

$$c(e' - d') + d(c' - e') + e(d' - c') = 0$$

Adding the last 4 equations together, we get

$$(a + c)(e' + f' - b' - d') + (b + d)(a' + c' - e' - f') + (e + f)(b' + d' - a' - c') = 0$$

which is equivalent to

$$g(i' - h') + h(g' - i') + i(h' - g') = 0$$

or G, H, I are co-linear as to be proved.

From the above we see that owing to Descartes geometry theorem-proving had entered a second stage of Descartes fashion which differs from the first stage of Euclidean fashion in that the characteristic feature (a) has to be replaced by (a') below:

(a') Proving is carried out by mere computations.

On the other hand the characteristic feature (b) remains unchanged. For each individual theorem one should seek a way of manipulating the algebraic equations involved to arrive at the goal, and this manipulation is peculiar to that theorem itself. Moreover, the discovery of such a way of proof is not at all a trivial one. Ingenuity and cleverness are still required in the algebraic manipulations as shown above in Gauss' proof of his theorem, though usually far less than what is required for a proof in Euclidean fashion. Furthermore, the computational work may be so enormous that it is out of control. These are perhaps the reasons why such powerful method of Descartes of geometry theorem-proving is not taught in courses on analytic geometry in high schools.

3. Stage of Geometry Theorem-Proving in Hilbertian Fashion

A further complete change in the fashion of geometry theorem-proving occured with Hilbert's classic "Grundlagen der Geometirie" appeared in 1899. For this Hilbert's method of theorem-proving let us take again Gauss' theorem G for illustration.

Consider Fig.1. again. Instead of considering the figure as drawn in one stroke, let us consider it as being drawn step by step by introducing the points and lines one-by-one in succession. Thus, we shall complete the figure in the following way.

First choose an arbitrary point A in the plane and take it as the origin. Through A choose now two arbitrary lines to be taken as the X-axis and Y-axis of an oblique coordinate system. Next choose an arbitrary point B on the Y-axis with coordinates $(0, u_1)$, u_1 being a parameter showing its arbitrariness. Choose an arbitrary point C in the plane with coordinates (u_2, u_3) and an arbitrary point D on the X-axis with coordinates $(u_4, 0)$. The $u's$ are again all parameters. Now join C, D by a line which meets the Y-axis at a point E. Thus the coordinates of E are not arbitrary. Let E be $(0, x_1)$ where x_1 is easily seen to be

$$x_1 = u_3 u_4/(u_4 - u_2). \tag{3.1}$$

Note that here we have implicitly assumed $u_4 - u_2 \neq 0$ implying that DC is not parallel to the Y-axis AB so that E is well defined and x_1 is well- determined by the above expression. Now join B, C by a line which meets the X-axis AD in a point F. The coordinates of F are not arbitrary. Let F be $(x_2, 0)$ then x_2 will be given by

$$x_2 = u_1 u_2/(u_1 - u_3). \tag{3.2}$$

As before we assume here $u_1 - u_3 \neq 0$ or BC is not parallel to the X-axis. Join now A, C and let their mid-point G be (x_3, x_4) so that

$$x_3 = \frac{1}{2}u_2, \quad x_4 = \frac{1}{2}u_3 \tag{3.3}$$

Join B, D *resp.* E, F and let their mid-points be *resp.* $H = (x_5, x_6)$ and $I = (x_7, x_8)$ so that

$$x_5 = \frac{1}{2}u_4, \quad x_6 = \frac{1}{2}u_1 \tag{3.4}$$

$$x_7 = \frac{1}{2}x_2, \quad x_8 = \frac{1}{2}x_1 \tag{3.5}$$

Equations (3.1) - (3.5) form the hypothesis system of theorem G in question. The conclusion of theorem G corresponds then to the equation below:

$$\begin{vmatrix} x_3 & x_4 & 1 \\ x_5 & x_6 & 1 \\ x_7 & x_8 & 1 \end{vmatrix} = 0 \tag{3.6}$$

For the proof let us start from the left-hand side of the conclusion (3.6) and eliminate $x_8, x_7, \ldots, x_2, x_1$, in succession by means of the hypothesis equations (3.5), ..., (3.2), (3.1). The computation runs as follows

$$\begin{vmatrix} x_3 & x_4 & 1 \\ x_5 & x_6 & 1 \\ x_7 & x_8 & 1 \end{vmatrix} \rightarrow \begin{vmatrix} x_3 & x_4 & 1 \\ x_5 & x_6 & 1 \\ \frac{1}{2}x_2 & \frac{1}{2}x_1 & 1 \end{vmatrix} \rightarrow \begin{vmatrix} x_3 & x_4 & 1 \\ \frac{1}{2}u_4 & \frac{1}{2}u_1 & 1 \\ \frac{1}{2}x_2 & \frac{1}{2}x_1 & 1 \end{vmatrix} \rightarrow \begin{vmatrix} \frac{1}{2}u_2 & \frac{1}{2}u_3 & 1 \\ \frac{1}{2}u_4 & \frac{1}{2}u_1 & 1 \\ \frac{1}{2}x_2 & \frac{1}{2}x_1 & 1 \end{vmatrix} \rightarrow \frac{1}{4}\begin{vmatrix} u_2 & u_3 & 1 \\ u_4 & u_1 & 1 \\ \frac{u_1u_2}{u_1-u_3} & \frac{u_3u_4}{u_4-u_2} & 1 \end{vmatrix} \rightarrow 0$$

The final result 0 shows that the theorem G is true provided that the denominator in (3.1) and (3.2) are not 0, which hold naturally as already pointed out. In comparison with previous proofs in Euclidean or Descartesian fashion we see that intelligence and cleverness play almost no role anymore. A layman with only rudiments of knowledge of geometry may well carry out the proof without much effort. What is all the more important is this: the above method of proof works not only for theorem G, but works as well for innumerous theorems of similar kind, in contrast to the Euclidean or Descartesian fashion proofs which work usually for the single theorem in question. Both the method and its universal applicability as described above are due to D. Hilbert. In fact, in the above mentioned classic work of Hilbert there was certain passage in Chap 6, section 35, which was turned into a theorem in later editions and is reproduced below:

Satz 62. Jeder reiner Schnittpunktsatz der in einer ebener Geometrie gilt, in der die Axiome ... und der Pascalschen Satz gültig sind, stellt sich durch Konstruktion geeigneter Hilfspunkte und Hilfsgeraden als eine Kombination endlich vieler Pascalsche Konfigurationen heraus.

The "reiner Schnittpunktsatz" in the above theorem was explained by Hilbert which we again reproduce as follows.

Unter einer reiner Schnittpunktsatz verstehen wir hier einen Satz, der eine Aussage über die vereinigte Lage von Punkten und Geraden und über die Parallelität von Geraden enthält, ohne weitere Beziehungen, wie etwa Kongruenz oder Senkrechtstehen, zu benutzen.

From this explanation we see clearly that the so-called "rein Schnittpunktsatz" includes a innumerous class of theorems, Desargues theorem, Pappus theorem, harmonic point-pair theorem, to mention just a few of them. Moreover, while the statement of theorem 62 is itself quite obscure, the passage before this statement makes clear that Hilbert had shown how any of the innumerous theorems in the whole class can be proved by this method and what the universal method is to achieve this. Instead of reproducing these passages again let us describe Hilbert's universal method in our own way as follows:

Proof Procedure:

Step 1. Define coordinate systems

Step 2. Introduce points and lines in succession

Step 3. Introduce parameters $u_1, u_2, \ldots$ and bounded coordinates $x_1, x_2, \ldots$ in succession so that hypothesis of the geometry theorem in question will be turned successively in the form below:

$$x_1 = P_1(u)/Q_1(u),\ x_2 = P_2(u, x_1)/Q_2(u, x_1) \cdots$$
$$x_r = P_r(u_1, x_1, \cdots, x_{r-1})/Q_r(u_1, x_1, \cdots, x_{r-1}) \tag{3.7}$$

In these equations all P_i, Q_i are polynomials in $u = (u_1, u_2, \cdots)$ and the previous bounded variables $x_1, \cdots, x_{i-1}$. Also convert the conclusion of the theorem in question into the form

$$R(u, x_1, \cdots, x_r) = 0 \tag{3.8}$$

Step 4. Eliminate $x_r, x_{r-1}, \cdots, x_2, x_1$ from R in succession by means of (3.7) such that

$$R(u, x_1, \cdots, x_r) \to R_1(u, x_1, \cdots, x_{r-1}) \to \cdots R_{r-1}(u, x_1) \to R_r(u)$$

Step 5. If $R_r(u) = 0$ then we conclude that the theorem in question is true provided that the conditions

$$Q_i(u, x_1 \cdots, x_{i-1}) \neq 0 \tag{3.9}$$

are observed. Otherwise we conclude that the theorem is not true at least under the conditions of equation (3.9) and is to be considered as not true at all.

We remark that for a theorem in the above-mentioned class to be true some restrictions like (3.9) are usually unavoidable. Accordingly we shall refer to such conditions as *non-negeneracy conditions.* To see its geometrical significance we may refer to the examples given in [WU2] or the example below.

Example *D. Desargues Theorem. Two triangles ABC and A′B′C′ are such that the 3 joining lines AA′, BB′, CC′ of the 3 pairs of corresponding vertices are concurrent in a point O. Suppose that two pairs of corresponding pairs of sides, say AB and A′B′ and also AC and A′C′ are mutually parallel to each other. Then the third pair of corresponding sides, viz BC and B′C′ , will also be parallel to each other ((Cf.)Fig. 3).*

For the proof let us follow the steps as indicated above. First choose the point O as the origin and two arbitrary lines through it as the oblique $X-$ and Y-axes, Choose now in succession arbitrary points $A = (u_1, 0)$. $A' = (u_2, 0)$ on X-axis, $B = (0, u_3)$ on Y-axis, and $C = (u_4, u_5)$ in the plane. Join AB and through A' draw a line parallel to AB meeting the Y-axis in a point $B' = (0, x_1)$, then $x_1 = u_2 u_3 / u_1$. Join AC, OC and through A' draw a line parallel to AC meeting OC in a point $C' = (x_2, x_3)$. It is easy to see that $x_2 = u_2 u_4 / u_1$ and $x_3 = u_5 x_2 / u_4$. The conclusion that BC is parallel to $B'C'$ corresponds to $u_4(x_3 - x_1) - (u_5 - u_3)x_2 = 0$. To prove it let us eliminate x_3, x_2, x_1 in succession from the left-hand side of the last equation. The computation gives

$$u_4(x_3 - x_1) - (u_5 - u_3)x_2 \to u_4\left(\frac{u_5 x_2}{u_4} - x_1\right) - x_2(u_5 - u_3)$$
$$= -u_4 x_1 + u_3 x_2 \to -u_4 x_1 + \frac{u_2 u_3 u_4}{u_1} \to 0$$

The Desargues theorem is thus seen to be true provided that the non-degeneracy conditions $u_1 \neq 0$ and $u_4 \neq 0$ are observed. The case $u_1 = 0$ will make A coincident with O and the theorem will become devoid of geometrical meaning. On the other hand the case $u_4 = 0$ will make C a point on the X-axis. According to the geometrical conditions in the theorem C' may become then an arbitrary point on the X-axis too. In that case the theorem is clearly not true as seen from Fig. 4. The non-degeneracy conditions $u_1 \neq 0$ and $u_4 \neq 0$ are thus indispensable for the theorem in question to be meaningful or to be true. We remark that this is the general situation of a geometry theorem and for further examples of such sort we may refer to e.g., [WU2].

From the above we see that Hilbert's method provided geometry theorem proving an entirely new 3rd stage of development. While the characteristic feature (a') of proving by computations remains the same as in the 2nd stage of Descartesian fashion, the feature (b) has been changed to (b') bleow.

(b') Proving is carried out not for theorems one by one, but class by class, i.e., a whole class of theorems will be proved by one and the same method.

4. Influences of Hilbertian Fashion of Geometry Theorem-Proving

Before entering upon discussions about the influences and further developments of Hilbertian fashion of geometry theorem-proving let us first make some remarks and comments about it.

Let us first remark that the proof of Gauss' theorem G as given in section 3 shows that the class of pure intersection theorems provable by Hilbert's method may be enlarged to include theorem G. In fact, a young mathematician D.M. Wang in China has pointed out that the class may be significantly enlarged to include theorems involving perpendicularities or even circles provided that the hypotheses can be so arranged that the bounded variables may be entered in succession as rational functions in the parameters as well as the previously introduced bounded variables. A lot of interesting non-trivial geometry theorems have indeed been proved by Wang in this manner, Cf e.g., [W-G] and also his Ph.D. thesis (unpublished).

On the other hand the Hilbert method has great limitations not only because it considers only theorems involving mainly incidence and parallelism properties alone (though it may be somewhat enlarged by Wang) but also because the theorems should be of a *constructive* character in the sense that points and lines, or rather bounded variables, should be introduced one-by-one in succession. That this may or may not easily be carried out, even if we restrict ourselves to theorems involving incidence and parallelism alone, may be seen from the following example due to Bokowski:

Example B. *Consider Fig.5. The numbers 1, 2, $\cdots$, 8 signify 8 points situated in the ordinary 3-space. A square in the figure with corners e.g.* $1, 2, 5, 6$ *will be said to be coplanar if the 4 points corresponding to* $1, 2, 5, 6$ *are coplanar. A conjecture of Bokowski says that if 7 of the 8 squares in the figure are coplanar, then the 8th one will also be coplanar.*

It is seen that the above conjecture is hardly constructive at least in appearance so the method of Hilbert can hardly be applied (though it turns out to be so at length). For another example see Ex. 7 of [WU2].

As a further remark let us consider the non-degeneracy conditions occuring in theorems mentioned in section 3. We see that all such non-degeneracy conditions are not clearly stated in statements of theorems to be proved. To see the grave consequence of this fact let us reproduce below some passage in the author's paper [WU3]:

> *... almost all theorems in ordinary geometry are only genetrically true. or true under some non-degeneracy conditions usually not explicitly described in the statement of the theorem. This fact is quite fatal to make the usual Euclidean fashion of proving theorems rigorous as one believes to be so, since it is difficult to state clearly the non-degeneracy conditions to be observed and to verify that the previous theorems to be applied in the proof of the theorems in question do not fall into the degeneracy cases under which the previous theorem to be applied might not be true.*

Though the class of theorems provable by the Hilbert method is quite limited the importance of the method is almost inestimable. Besides the novelty of proving the theorems not one-by-one but by a single universal method for those in a whole class, the proving method itself is a novel one: it is mechanical in character in that one may just follow the instructions step-by-step in a mechanical way in order to arrive at the final conclusion of the theorem in question. For the recognition of the merit of the Hilbert method let us reproduce a passage of the author's paper [WU 2] as given below:

> *The great merit of Hilbert's classic "Grundlagen der Geometre" of 1899 is universally recognized as being representative for axiomatization of mathematics, laying in particular a rigorous foundation of the Euclidean geometry. However, another great merit (perhaps greater in the opinion of the present author) of this classic seems hardly to be noticed up to the present. In fact, this classic is also representative for the mechanization of geometry, showing clearly at the same time the way to achieve it.*

We remark in passing that the saying "rigorous foundation" may be of some doubt, in view of the occurrence of non-degeneracy conditions as mentioned above.

Owing to the mechanical character of the Hilbert method we shall reformulate the Satz 62 of Hilbert's classic as the following

Hilbert Mechanization Theorem. The class of planar pure intersection theorems is mechanizable.

This concept of mechanizability has in fact prevailed in the whole of mathematics far beyond the limited class of theorems above, but under a different terminology of decidability which we give the definition below.

Definition *A class or domain of mathematical theorems (resp. mathematical problems), is said to be decidable if there is an algorithmic procedure which if followed step by step one can see in a finite number of steps whether any theorem (resp. any problem) in the class or domain can be proved to be true (resp. can be solved) or not and gives the proof (resp. the solution) in the affirmative case. Instead of "decidable" we prefer to use "mechanizable" in what follows. In an abuse of language we shall also say that a domain of mathematics is mechanizable if this is true for the whole class of theorems (or problems) in that domain.*

Under the guiding thoughts of Hilbert a new discipline of mathematics, viz. mathematical logic, has appeared and flourished in this century. One of the main topics of study of this discipline is just the study of mechanizability (i.e., decidability) of various domains of mathematics. Among the topmost results thus attained let us just cite the following few.

M1. The domain of arithmetic involving addition alone is mechanizable, while that of arithmetic involving both addition and multiplication is non-mechanizable.

M2. The word problem of the class of all groups is non-mechanizable, while the word problem of certain smaller class of groups, e.g., that of free groups, is mechanizable.

M3. Hilbert's 10th problem has been solved in the negative in 1970, viz. the solution of the whole class of diophantine equations is non-mechanizable, while the restriction to the much smaller class of linear diophantine equations has been known to be mechanizable since early in the 19th century, and which may even be traced back to China and India in ancient times.

M4. A remarkable theorem about mechanical geometry theorem proving was discovered by Tarski in 1950 which will be called the

Tarski Mechanization Theorem. Elementary geometry in the ordinary sense is mechanizable.

It is clear that the Hilbert Mechanization Theorem is a very particular case of Tarski's in that it deals with a very small domain contained in the whole domain considered by Tarski. On the other hand, little is known about the mechanizability of other kinds of geometry.

From the above examples M1-M4 we may deduce the following conclusions:

If a domain of mathematical theorems or problems is too vast, then it may be non-mechanizable.

If a domain of mathematical theorems or problems is too small, then it may be of little or no mathematical interest, even though it is mechanizable.

These indications serve as clues to the search of domains of mathematics to be studied:

1. The domain should be sufficiently small to be mechanizable.
2. The domain should be sufficiently large to be of mathematical interest.

These will serve as guidelines for our future study.

5. Stage of Geometry Theorem-Proving by Computer

Geometry theorem-proving in Hilbertian fashion has received an immense impetus since the appearance of computers in the forties. In fact, the very mechanical character of Hilbert method of proving permits us to turn the proving procedure directly into a program which can then be run on the computer to see whether the theorem in question is true or not. The running computations of the computer is then itself a proof or a disproof of the theorem. With computers as powerful tools, geometry theorem-proving entered a new 4th stage of proving by computer. In this stage characteristic feature ($a\prime$) is further strengthened to

($a\prime$) Proving is carried out by computations on a *computer*.

The characteristic feature (*b*′) of proving class by class will remain unchanged. However, the class will no more be restricted to geometry alone and may be any domain of mathematical theorems or even problems.

With these understood the fifties saw a period of earnest trials in proving theorems, geometrical ones or not by computer. A breakthrough occured eventually in the very beginning of the sixties. Wang Hao, a Chinese mathematical logician, proved on the IBM 704 computer approximately two-hundred theorems of propositional calculus in barely 3 minutes. The pioneering achievement was rightly cheered as "seven flies in one blow" and Wang Hao was accordingly awarded the first mechanical or automated theorem-proving prize in 1983.

The development of the stage of theorem-proving in Euclidean fashion to the present stage of proving by computer may be synthesized in the vivid expression of Wang Hao: "to trade effectively quantitative complexity with qualitative difficulty." While qualitative difficulty of geometry theorem-proving had been overcome for the most through the work of Descartes and Hilbert, the quantitative complexity will be seen to be overcome by the use of computers. The success of Wang Hao thus induced great efforts in seeking for more powerful methods of proving more profound theorems by computer. Through the sixties and onwards there appeared thus various kinds of resolution and other methods for this purpose, for a summary of which we refer to [B-L].

However, all the above achievements, though of great theoretical importance, seem to be of little interest for the taste of a working mathematician. From the point of view of a mathematician, it would be much more interesting to see the outcome of a computer implementation of Tarski's method in his mechanization theorem.

The results seem to be somewhat discouraging. The method of Tarski is too complicated to be efficient so that actually no theorems of some geometrical interest have ever been proven even with the large present-day computers.

The reason for the lack of success of Tarski's method may be indicated in the guidelines at the end of the last section. Though the domain of theorems treated by Hilbert's mechanization theorem seems to be a little too narrow to have great geometrical interest, the domain of the theorems considered by Tarski is surely too large to render the method efficient enough in proving interesting theorems.

To rescue this situtation we have to seek, therefore, some domain midway between those of Hilbert and Tarski. Such a domain should be larger than that of Hilbert to encompass much more interesting theorems in elementary geometry, at the same time it should be smaller than that of Tarski to be efficient enough to prove still a large number of interesting theorems without losing too many. In short, we have to be modest. We have to be less ambitious. We have to retreat in order to gain without too much loss.

This was what a small group called the "mathematics-mechanization group", a group of Chinese mathematicians in the Institute of Systems Science belonging to the Academy of Sciences of China, has done since the year 1977 [WU1]. We shall descrìbe the activities of this small group in more detail in later sections [WU1-6].

6. Mechanical Geometry Theorem Proving in China

Resolution methods for mechanical theorem-proving have prevailed in China since the sixties. Among the leaders we may mention in particular Prof. S. H. Wang, a reknown algebraist in China. However, mechanical theorem-proving applied to geometries began only late in the seventies with the appearance of a paper by Wu, [Wu 1]. The discovery of such a method of mechanical geometry theorem-proving occurred in the beginning of 1977. In fact Tarski's method consists of reducing the theorems in ordinary elementary geometry to algebraic forms with both hypotheses and conclusions expressed as conjunctions and disjunctions of polynomial equations or inequalities. We may put Tarski's formulation in the following concise form:

$$\left\{\vee, \wedge, f_i(x) \overset{\geq}{\underset{\leq}{=}} 0\right\} \rightarrow \left\{\vee, \wedge\ g_j(x) \overset{\geq}{\underset{\leq}{=}} 0\right\} \tag{6.1}$$

As the ordinary Euclidean geometry is real in character and order relations between the geometrical entities play an important role which correspond to inequalities in algebraic formulations so Tarski had to apply Sturm's theorem to deal with such situations. On the other hand, in Hilbert's method of mechanization only equations will be involved and may be put in the following concise form:

$$\{x_i = P_i(u, x_1, \cdots, x_{i-1})/Q_i(u, x_1 \cdots, x_{i-1}), i = 1, \cdots, n\} \to \{g(u, x_1, \cdots, x_n) = 0\} \quad (6.2)$$

Here P_i, Q_i, g are all polynomials and the hypothesis equations are linear in the bounded variables x_i in succession. The linearity of these equations imposes a great restriction to the domain of theorems which can be considered so that theorems about circles are largely excluded. On the other hand, Tarski's method as well as the later improvements are too complicated to be efficient.

As pointed out at the end of last section, a remedy to meet such a situation is to seek out some domain of geometry theorems midway between those of Hilbert and Tarski. Now the order relations are somewhat peculiar to ordinary elementary geometry. They are not so significant as in the case of differential geometry. They may even be utterly neglected as in the case of modern algebraic geometry for which the basic field is usually that of the complex numbers or one of characteristic 0. It seems therefore adequate in first restricting the domain of Tarski's method to one which involves equations alone but no inequalities at all. On the other hand, in order to include theorems involving circles one has to enlarge the domain of Hilbert to one for which the equations are not merely linear ones, but at least quadratic ones in the leading bounded variables. We are thus led to the consideration of such a domain of theorems which algebraically may be put in the following concise form:

$$\{h_i(u, x_1, \cdots, x_i) = 0, i = 1, \cdots, r\} \to \{g(u, x_1, \cdots, x_r) = 0\} \quad (6.3)$$

Here u stands for a set of parameters $u = (u_1, \cdots, u_d)$ and h_i, g are all polynomials. To see how such a theorem may be proved let us write h_i more precisely below:

$$h_i = I_i x_i^{d_i} + \text{lower degree terms in } x_i \quad (6.4)$$

In the expression the leading coefficient I_i as well as other coefficients are themselves polynomials but in the variables u and $x_1, \cdots, x_{i-1}$ alone. Let us now divide the conclusion poylnomial g by the hypothesis *polynomials* h_i in succession in the order $h_r, h_{r-1}, \cdots, h_2, h_1$. This will give us a set of identities in polynomials with successive remainders $R_{r-1}, \cdots R_0$ as shown below:

$$I_r^{s_r} g = Q_r h_r + R_{r-1},$$

$$I_{r-1}^{s_{r-1}} R_{r-1} = Q_{r-1} h_{r-1} + R_{r-2},$$

$$\cdots$$

$$I_1^{r_1} R_1 = Q_1 h_1 + R_0.$$

Combining these identities we will get one of the form

$$(I_1^{s_1} \cdots I_r^{s_r}) g \equiv R_0 \; mod(h_1, \cdots, h_r) \tag{6.5}$$

The final poylnomial R_0 will have its degree in $y_1, \cdots, y_r$ less than $d_1, \cdots, d_r$ *resp.* and will be called the *remainder* of the poylnomial g w.r.t. the set $HYP = \{h_1, \cdots, h_r\}$, to be denoted as

$$R_0 = Remdr(g/HYP)$$

Write $CONC$ for g so that the theorem in question, say T may be put in the form $T = \{HYP, CONC\}$. Then trivially we get the following:

Theorem I. *Consider a theorem $T = \{HYP, CONC\}$ with $HYP = \{h_1, \cdots, h_r\}$ and $CONC = g$ as above. Suppose that for the remainder we have*

$$Remdr(CONC/HYP) = 0 \tag{6.6}$$

Then from $HYP = 0$ it would follow $CONC = 0$, provided that $I_i \neq 0, i = 1, \cdots, r$. In other words, the theorem T is then generically true in the sense that it will be true under some non-degeneracy conditions

$$I_i \neq 0, i = 1, \cdots, r \tag{6.7}$$

We have already seen from examples for Hilbert Mechanization Theorem that such non-degeneracy conditions are usually unavoidable. Now the above Theorem I furnishes only a sufficient condition for a theorem T of above form to be true. However it has an important converse as follows.

Up to now we have not made precise the coefficient field of poylnomials (to be abbreviated as *pols* in what follows) involved. This is usually the field of rational numbers, to be representable on a computer. Let us admit this to be so and denote it by K. Let $K_0 = K(u_1, \cdots, u_d)$ be the transcendental extension field of K formed by adjoining to it the parameters u_i. Suppose now h_1 is irreducible in K_0, considered as a poylnomial or *pol* in $K_0[x_1]$. Form the algebraic extension field $K_1 = K_0(x_1)$ by adjoining x_1 to K_0 by means of the equation $h_1 = 0$. Suppose h_2 is irreducible in K_1, considered as a *pol* in $K_1[x_2]$. Adjoin now x_2 to K_1 by means of $h_2 = 0$ to form a new algebraic extension field $K_2 = K_1(x_1)$. Suppose it may be continued up to h_r so that we have finally $K_r = K_{r-1}(x_r)$. Then we say simply that HYP is an irreducible set. As a complement to Theorem I we have the following:

Theorem $I\prime$: Suppose given a theorem $T = \{HYP, CONC\}$ as above for which the hypothesis set HYP is irreducible. Then for T to be generically true under the non-degeneracy conditions (6.7) it is necessary that (6.6) should hold true.

To see the strength of the above Theorem I let us consider as an example the following celebrated

Feuerbach Theorem F. The nine-point circle of a triangle is tangent to all the four inscribed and es-cribed circles of that triangle.

We recall that the so-called nine-point circle is one which passes through the mid-points of the 3 sides, the 3 feet of altitudes, and the mid-points between the ortho-center and the 3 vertices of the triangle. This nine-point circle theorem is itself very easily proved by means of the method in Theorem I. Admitting this let us proceed to prove the much more delicate theorem of Feuerbach as follows.

The first step is to choose a convenient coordinate system. If we choose at first the

3 vertices A, B, C of the triangle with coordinates in parameters as in the case of Hilbert Mechanization Theorem then there will be an ambiguity for the coordinates of the in- or es-center, say I. To avoid this let us thus choose first the points I and A, B with coordinates

$$I = (0, u_1), A = (u_2, 0), B = (u_3, 0).$$

The third vertex C will then be well determined as the intersection point of the reflection lines of the side AB with respect to the lines AI and BI, provided that the angle AIB is not a right angle in which case the two reflection lines will become parallel. Let $C = (x_1, x_2)$ and the nine-point circle center be $N = (x_3, x_4)$. Let the radius of the nine-point circle and the distance between N and I be x_5 and x_6, *resp.* Note that x_5 and x_6, as the radius of a circle or a distance should both be positive. As positiveness or negativeness are not to be considered in our method of Theorem I, there will arise some ambiguities in the formulation. However, this will not cause many difficulties, at least in the present case, as we shall see below.

It is easy to turn the geometrical conditions in the theorem into equations $HYP = 0$ for the hypothesis and $CONC = 0$ for the conclusion. The hypothesis set HYP will be consists of 6 pols $h_1 \cdots, h_6$ as follows:

$$h_1 = (u_1^2 + u_2u_3)x_1 - u_1^2(u_2 + u_3),$$
$$h_2 = (u_1^2 + u_2u_3)x_2 - 2u_1u_2u_3,$$
$$h_3 = 4x_3 - (u_2 + u_3 + 2x_1),$$
$$h_4 = 4x_2x_4 - x_2^2 - (u_1 - x_1)(4x_3 - 2u_2 - u_3 - x_1),$$
$$h_5 = 4x_5^2 - (2x_3 - u_2 - u_3)^2 - 4x_4^2,$$
$$h_6 = x_6^2 - x_3^2 - (x_4 - u_1)^2.$$

We remark that each h_i with $i = 1, \cdots, 4$ is linear in the corresponding x_i as in the case of Hilbert's Mechanization Theorem. On the other hand h_5 and h_6 are quadratic in x_5 and

x_6, and cannot be written in a form linear in x_5 or x_6, which reflects the fact that the positiveness or negativeness of the radius x_5, x_6 is indistinguishable in our formulation. For the same reason the conclusion to be proved is to be put in the form $x_6 = \pm x_5 \pm u_1$. We leave the ambiguities of sign + or - undecided reflecting the geometrical fact that we neglect the manner of contact, inner-wise or outer-wise, of the nine-point circle and the in- or es-cribed circle. In any case the theorem will be proved if we can prove that

$$(x_6 + x_5 + u_1)(x_6 - x_5 - u_1)(x_6 + x_5 - u_1)(x_6 - x_5 + u_1) = 0$$

leaving aside the manner of contact. The conclusion pol $CONC$ can thus be taken to be the left hand side of the last equation, viz.

$$CONC = x_6^4 + x_5^4 + u_1^4 - 2x_6^2x_5^2 - 2x_6^2u_1^2 - 2x_5^2u_1^2$$

Form now the remainder of $CONC$ by dividing it successively by $h_6, h_5, \cdots, h_1$ to remove $x_6, x_5, \cdots, x_1$ as much as possible. We find that

$$Remdr(CONC/HYP) = 0 \tag{6.8}$$

or more precisely

$$4^{s_1}x_1^{s_2}(u_1^2 + u_2u_3)^{s_3} \cdot CONC \equiv 0 \bmod (HYP), \tag{6.9}$$

in which $4, x_1, u_1^2 + u_2u_3$ are the leading coeffcients of h_i and s_i are integers ≥ 0.

The identities (6.8) or (6.9) show that the Feuerbach theorem is true under the non-degeneracy conditions $x_2 \neq 0$ and $u_1^2 + u_2u_3 \neq 0$. The last condition means that the angle AIB should not be a right angle, which is naturally observed so it is not of consequence.

For the above example it has been shown that the method described in Theorem I, simple and trivial as it is, is powerful enough to prove quite non-trivial theorems. In fact, hundreds of non-trivial theorems in elementary geometry have been proved and even discovered both in China and abroad, USA in particular. Moreover, S.C. Chou at University of Texas at Austin had made a gigantic scale of experiments about the method, and a book of Chou has been published in recent years which contains such mechanical proofs of more than 500 difficult geometry theorems, [CH] and also [W-G].

7. Some General Principles Underlying Mechanical Geometry Theorem-Proving

The mechanical method of geometry theorem-proving as described in Theorem I of section 6, though efficient and fruitful enough, is quite incomplete in itself. Besides the fact that we have utterly left aside the theorems involving order relations in the geometrical sense or inequalities in the algebraic sense, we have restricted ourselves to theorems for which the hypotheses poylnomials have to be put in the form of $h_i = 0$ in (6.3). This is surely not always the case. Moreover, hypotheses put in the above form may be such that it is impossible to prove the theorems in question to be true in the above way, though they are well-known to be true. This may be due to some ambiguities in the statements of the theorems which occur quite often as in the case of above Feuerbach Theorem. A further shortcoming is this: while numerous theorems will involve trigonometric, or even hyperbolic functions as in the case of non-Euclidean geometries, we have restricted ourselves to theorems which will be expressed solely in the form of polynomial equations.

Let us try to answer the last question first.

For an expression involving some trigonometric functions like $\sin u, \cos u$ we may replace them by some new variables x, y connected by an algebraic relation $x^2 + y^2 = 1$. An alternative way is to introduce a new variable z standing for $\tan \frac{1}{2}u$ and replace $\sin u, \cos u$ by the algebraic expressions $2z/(1+z^2)$ and $(1-z^2)/(1+z^2)$, respectively. Such rationalization procedures in introducing new variables and new algebraic relations may quite often, though not always, be enough to attain our goal of expressing the hypothesis and conclusion of theorems to be considered in the form of polynomial equations. There are however, much deeper reasons that one may always attain our goal of restricting ourselves to polynomial equations alone. This may be seen as follows.

Let us look at the Hilbert's classic "Grundlagen der Geometrie" at greater length

In his classic Hilbert defined ordinary Euclidean geometry as the system of a set of undefined geometrical entities, points, lines, etc., a set of undefined geometrical relations between such entities, incidence, parallelism, etc., and finally a set of axioms connecting these entities and relations. Hilbert then proceeded to show how, starting from such a system and using logical reasoning alone, certain number systems may be introduced which is in the present case a real closed field intrinsic to that geometry. With the aid of such a number system a coordinate system is then introduced in the usual manner. We may schematize the above procedure in the diagram below:

$$\text{Geometry} \rightarrow \text{Number System} \rightarrow \text{Coordinate System} \tag{7.1}$$

It turns out that in terms of such coordinate systems all the geometrical relations which form one of the fundamental sets defining that geometry, will be expressable in the form of polynomial equations.

Now this is not an isolated fact true for ordinary Euclidean geometry alone.

In a booklet about mechanical geometry theorem-proving written in Chinese and published in 1984, [WU5], the present author showed that a hierarchy of geometries may be defined in well-choosing and well-arranging the fundamental sets of geometrical entities, geometrical relations and geometrical axioms from those considered by Hilbert. Other kinds of geometries, affine geometry, projective geometry, non-Euclidean geometries both elliptic and hyperbolic, also line geometry and circle geometries may also be defined in a similar way. For all such geometries one may introduce number systems and then coordinate systems as indicated in the scheme (Fig.7.1). It turns out that for each such geometry, provided that the order relations are discarded and a certain axiom of Pascalian type in the sense of Hilbert is admitted, all the fundamental relations meaningful for that geometry can be expressed in the form of polynomial equations.

It follows that the restriction to polynomial equations is not a serious one since the above-mentioned geometries possessing this property encompass actually most of the geometries of interest.

Come now to the question for which the hypothesis sets of polynomials, or *pols* for short, are put in the form as exhibited in (6.3)

Consider thus a geometry theorem $T = \{HYP, CONC\}$ with HYP the hypothesis polynomial set or *polset* for short and $CONC$ the conclusion pol. We suppose that all these pols are in the ring $K[x_1, \cdots, x_r]$ with K a field of characteristic 0 intrinsic to the geometry in question, but otherwise arbitrary.

Let us call a zero of a pol in $K[x_1, \cdots, x_r]$ any point in the n-dimensional affine space K'^n with K' any extension field of K such that P becomes 0 when each x_i is replaced by the i-th coordinate of that point. For a polset PS in $K[x_1, \cdots, x_r]$ a zero is then one which is a common zero of all the pols in PS. The totality of all zeros of a pol P *resp.* a polset PS will be denoted by $Zero(P)$ and $Zero\,(PS)$ *resp.* The totality of zeros of PS which are not zeros of some other pol G will then be denoted by $Zero\,(PS/G)$.

Let the polset PS be now our hypothesis set HYP. A zero of HYP is then nothing else but a geometrical configuration verifying the hypothesis of the geometry theorem in question, which may eventually be in some imagined extended space, via, K'^n for some extension field K'. The conclusion of the theorem will then be true for that geometrical configuration corresponding to a zero of HYP if and only if this zero is also one of $CONC$. In this sense proving of a theorem T may be regarded as the determination of such a subset S of $Zero(HYP)$ such that $CONC = 0$ for all points in S or $Zero(S) \subset Zero(CONC)$. From examples in previous section we see that in general S can only be a proper part of $Zero(HYP)$.

From such a formulation it is clear that if $Zero(HYP)$ is known then the determination of such a subset S will be relatively easily done. The problem of geometry theorem-proving, at least in the cases considered here, may thus be reduced to the problem of equations-solving, via. the determination of $Zero(HYP)$ or solving the equations $HYP = 0$. Thus we arrive at the general principle in considering geometry theorem-proving as an application of polynomial equations solving.

Consider an arbitrary polset PS in the ring $K[x_1, \cdots, x_r]$ with K a field of characteristic 0. Before coming to equations $PS = 0$ in the general case let us first consider, in view of (6.3), the case of polsets AS in some particular forms as follows.

Suppose that after some permutation the set of variables $(x_1, \cdots, x_n)$ may be put in the form $(u_1, \cdots, u_d, y_1, \cdots, y_r)$ with $d + r = n$ so that the pols in the set AS may be arranged as $A_1, \cdots,\ A_r$ in the form below:

$$\begin{aligned} A_1 &= I_1 y_1^{d_1} + \text{ lower degree terms in } y_1 \in K[u_1, y_1], \\ A_2 &= I_2 y_2^{d_2} + \text{ lower degree terms in } y_2 \in K[u_1, y_1, y_2], \\ &\cdots \\ A_r &= I_r y_r^{d_r} + \text{ lower degree terms in } y_r \in K[y, y_1, \cdots, y_r] \end{aligned}$$

We require also that for each i and any $j < i$, the degree in y_j of the leading coefficient I_i of A_i will be less than d_j. In this case, AS will be said to be an *Ascending Set* (abbr. asc-set). If we neglect the last condition about the degree then we shall call such a set a *Triangulated Set.* Furthermore, a polset consisting of a single pol which is a non-zero constant in K will also be considered as a *trivial* asc-set or triangulated set. It is clear that for an asc-set or a triangulated set AS the zero-set $Zero(AS)$ may be regarded as well-determined, which is simply the empty set in case of trivial ones.

The following fundamental definition is now crucial to our method of equations solving.

Definition *A polset CS is said to be a Characteristic Set (abbr. char-set) of a polset PS if it verifies the following conditions*

(C1) CS is an asc-set (or a triangulated set)

(C2) $Zero(PS) \subset Zero(CS)$.

(C3) $Remdr(P/CS) = 0$ for any P in PS.

The following fundamental lemma is now easily proved.

Lemma *If CS is a char-set of a polset PS, then*

$$Zero(PS) = Zero(CS/J) + SUM_i\ Zero(PS + \{I_i\}). \qquad (I)$$

in which I_i are leading coefficients of pols in CS and J is the product of all these I_i.

By making use of theory of J.F. Ritt as exhibited in his books [R1], [R2] we may derive an algorithm which permits us to determine from a given polset PS a char-set CS in a finite number of steps. Applying this to the hypothesis set HYP of a theorem $T = \{HYP, CONC\}$, we will get the following (MTP stands for mechanical theorem-proving)

(Special) MPT-Principle. Let CS be a char-set of the hypothesis set HYP with I_i as the leading coefficients of its pols. If

$$Remdr(CONC/CS) = 0$$

then the theorem $T = \{HYP, CONC\}$ is generically true under the non-degeneracy conditions

$$I_i \neq 0.$$

For each polset $PS + \{I_i\}$ in (I) it is clear that

$$Zero(PS + \{I_i\}) = Zero(PS + CS + \{I_i\}$$

Let us now treat each polset $PS + CS + \{I_i\}$ as the original polset PS to find a char-set and proceeding in the same way as far as possible we shall terminate in a finite number of steps and arrive finally at the following:

Zero Decomposition Theorem. There is an algorithm which given any polset PS determines for in a finite number of steps a finite number of asc-sets or triangulated sets AS_i such that

$$Zero(PS) = SUM_i \; Zero(AS_i/J_i), \qquad (II)$$

in which each J_i is the product of all leading coefficients of pols in the corresponding asc-sets or triangulated set AS_i.

Applying the above decomposition formula (II) to the case $PS = HYP$, the hypothesis set of a theorem T, we get then the following:

(General) MTP-Principle. For a theorem $T = \{HYP, CONC\}$ let AS_i be a set of asc-sets or triangulated sets determined from HYP such that

$$Zero(HYP) = SUM_i \; Zero(AS_i/J_i)$$

If for some i we have

$$Remdr(CONC/AS_i) = 0,$$

then T will be true on that part $Zero(AS_i/J_i)$ of $Zero(HYP)$ and J_i may be considered to be the corresponding non-degeneracy conditions.

In case some of AS_i in the decomposition (II) is not irreducible we may factor it, replace it by some other polsets and proceed as before. In this way we will finally arrive at some decomposition finer than (II), viz.

$$Zero(PS) = SUM_k\ Zero(IRR_k/J_k)$$

in which each IRR_k is an irreducible asc-set. For each k let ID_k be the ideal of all pols for which the remainders with respect to IRR_k are equal to 0. The algebraic variety of all zeros of the ideal ID_k is in the long run completely determined by the irreducible asc-set IRR_k and may be legitimately denoted as $Var[IRR_k]$. Note that $Zero[IRR_k]$ is always contained in $VAR(IRR_k)$ but in general not identical to it. For this reason we have denoted this variety through the use of brackets [ ] instead of parentheses () to distinguish it from the variety $Var(IRR_k)$, which is the one associated with the ideal with IRR_k as a basis. From the above we see that

$$Zero(PS) \subset SUM_k\ Var[IRR_k]\,.$$

It turns out that this inclusion is in fact an identity and we have the following:

Variety Decomposition Theorem. There is an algorithm which for any polset PS determines a finite number of irreducible asc-sets IRR_k such that

$$Zero(PS) = SUM_k\ Var[IRR_k]\,. \qquad (III)$$

Applying this theorem to theorm-proving we get the following:

(Complete) MTP - Principle. For a theorem $T = \{HYP, CONC\}$ let IRR_k be a set of irreducible asc-sets determined from HYP such that

$$Zero(HYP) = SUM_k\ Var[IRR_k]$$

Then the theorem T will be true on the whole of the variety $Var[IRR_k]$ if and only if

$$Remdr(CONC/IRR_k) = 0\,.$$

The final MTP-Principle is said to be complete in the sense that it gives necessary and sufficient conditions for a theorem to be true on each of the whole irreducible components into which the total variety of geometrical configurations verifying the hypothesis is to be decomposed. In comparison with the other two MTP-Princples which furnish only sufficient conditions for theorems to be true, we see that the introduction of non-degeneracy conditions is in reality some provisional artifice required to avoid excessively complicated computations for a decomposition of the relevant variety into irreducible components. Such a decomposition requires factorization of pols in an algebraic field, which is recognized to be a highly difficult task. However, for the case of ordinary Euclidean geometry such factorization may sometimes be achieved in a quite simple manner from geometrical considerations by introducing so-called oriented lines and oriented circles. In this way a completely mechanical proving of quite intricate theorems may be achieved. For more details we shall refer to the relevant paper [WU6].

8. Conclusion

In the preceding sections we have described how the theorem-proving has evolved and developed since the time of Euclid up to the present computer age. We may announce in view of the present status the following program as our ultimate goal:

Mathematics-Mechanization Program. Cover as much as possible the whole of mathematics by domains which are sufficiently small to be mechanizable and at the same time sufficiently large to have real mathematical interest.

We have some achievements in discovering some such domains in elementary geometries. We have discovered that the same is true for some domains of differential geometries. We hope that achievements may be attained also in other domains of mathematics. Of course, to have real significant advances in such vast domains of mathematics it will require the enormous efforts of an innumerable number of an mathematicians for centuries, not merely for years, to come.

REFERENCES

[B-L] Bledsoe, W. W. & Loveland, D. W. (eds), *Automated theorem proving: after 25 years*, Amer. Mathematics Society (1984).

[CH] Chou, S. C., *Mechanical geometry theorem proving*, Reidel (1988).

[C-L] Chang, C. L. & Lee, C.T., *Symbolic Logic and Mechanical theorem Proving*, Acad. Press (1973)

[D] Descartes, R., *La géomtrie*, Paris (1637).

[E] Euclid, *Elements*, 3c B.C.

[G] Gauss, K. F., *Bestimmung der grössten Ellipse walche die vier Seiten einen gegebenen Vierecks berührt*, in Werke, Bd 4 (1882) 385-392.

[H] Hilbert, D., *Grundlagen der Geometrie*, (1899).

[R1] Ritt, J. F. , *Differential equations from the algebraic standpoint*, Amer. Math. Soc. (1932).

[R2] Ritt, J. F., *Differential algebra*, Amer. Math. Soc. (1950).

[T] Tarski, A., *A decision method for elementary algebra and geometry*, Berkeley (1950).

[W-G] Wang, D. M. & Gao, X.S., *Geometry theorems proved mechanically using Wu's method*, Math-Mech. Res. Preprints, Inst. Systems Science, No. 2 (1987) 75-105.

[WU1] Wu Wen-tsün, *On the decision problem and the mechanization of theorems in elementary geometry*, Scientia Sinica 21(1978) 159-172, Re-publication in [B-L], 213-234.

[WU2] Wu Wen-tsün, *Toward mechanization of geometry-some comments on Hilbert's "Grundlagen der Geometrie"*, Acta Math. Scientia, 2 (1982) 125-188

[WU3] Wu Wen-tsün, *Some remarks on mechanical theorem proving in elementary geometry*, Acta Math. Scientia, 3 (1983) 357-360.

[WU4] Wu Wen-tsün, *Basic principles of mechanical theorem proving in elementary geometries.* J. Sys. Sci. Math. Scis., 4 (1984) 207-235. Republished in J. Automated Reasoning, 2 (1986) 221-252.

[WU5] Wu Wen-tsün, *Basic principles of mechanical theorem proving in geometries (Part of elementary geometries)*, Science Press, Beijing (1984) (in Chinese).

[WU6] Wu Wen-tsün, *On reducibility problems in mechanical theorem proving of elementary geometries*, Ch. Quart. J. Math., 2 (1989) 1-20, or Math.-Mech. Res. Preprints, Inst. Sys. Sci., No. 2 (1989) 18-36.

Progress in Computer Science and Applied Logic

Progress in Computer Science and Applied Logic is a series that focuses on scientific work of interest to both logicians and computer scientists. Thus both applications of mathematical logic will be topics of interest. An additional area of interest is the foundations of computer science.

The series (previously known as *Progress in Computer Science*) publishes research monographs, graduate texts, polished lectures from seminars and lecture series, and proceedings of focused conferences in the above fields of interest. We encourage preparation of manuscripts in some form of TeX for delivery in camera-ready copy, which leads to rapid publication, or in electronic form for interfacing with laser printers or typesetters.

Proposals should be sent directly to the editors or to:
Birkhäuser Boston, 675 Massachusetts Ave., Cambridge, MA 02139

Progress in Computer Science and Applied Logic

PCS 1 Mathematics for the Analysis of Algorithms, 3rd Edition
Daniel H. Greene & Donald E. Knuth

PCS 2 Applied Probability--Computer Science: The Interface, Volume I
Edited by Ralph L. Disney & Teunis J. Ott

PCS 3 Applied Probability--Computer Science: The Interface, Volume II
Edited by Ralph L. Disney & Teunis J. Ott

PCS 4 Notes on Introductory Combinatorics
George Pólya, Robert E. Tarjan, & Donald R. Woods

PCS 5 The Evolution of Programs
Nachum Dershowitz

PCS 6 Lecture Notes on Bucket Algorithms
Luc Devroye

PCS 7 Real-Time Control of Walking
Marc D. Donner

PCS 8 Logic for Computer Scientists
Uwe Schöning

PCS 9 Feasible Mathematics
Edited by Samuel R. Buss & Philip J. Scott

PCS 10 Graph-Based Proof Procedures for Horn Clauses
Stan Raatz

PCS 11 A Proof Theory of General Unification
Wayne Snyder

PCS 12 Logical Methods: In Honor of Anil Nerode's Sixtieth Birthday
Edited by John E. Crossley, Jeffrey B. Remmel, Richard A. Shore, & Moss E. Sweedler